Lecture Notes in Mathematics

Edited by A. Dold, Heidelberg and B. Eckmann, Zürich

Series: Dept. of Mathematics, Univ. of Maryland, College Park
Adviser: J. K. Goldhaber

352

John D. Fay
University of Maryland, College Park, MD/USA

Theta Functions on Riemann Surfaces

Springer-Verlag
Berlin · Heidelberg · New York 1973

AMS Subject Classifications (1970): 30-02, 30 A 48, 30 A 58

ISBN 3-540-06517-2 Springer-Verlag Berlin · Heidelberg · New York
ISBN 0-387-06517-2 Springer- Verlag New York · Heidelberg · Berlin

Offsetdruck: Julius Beltz, Hemsbach/Bergstr.

Preface

These notes present new as well as classical results from the theory of theta functions on Riemann surfaces, a subject of renewed interest in recent years - [5], [10]. Topics discussed here include: the relations between theta functions and Abelian differentials, theta functions on degenerate Riemann surfaces, Schottky relations for surfaces of special moduli, and theta functions on finite bordered Riemann surfaces.

I wish to express sincere thanks to Prof. Lars V. Ahlfors for his constant help and encouragement over many years, and to Prof. David Mumford for generous assistance at several points in this work.

Research for these notes was supported by the National Science Foundation.

Table of Contents

I. Riemann's Theta Function

A variation of the classical Krazer notation [19] will be used in these notes: a principally polarized complex Abelian variety $\mathcal{A}$ of dimension g will be written as $\mathcal{A} = \mathbb{C}^g/\Gamma$, where Γ is the lattice in $\mathbb{C}^g$ generated by the columns of the $g \times 2g$ matrix $(2\pi i I, \tau)$ with I the identity $g \times g$ matrix and τ a symmetric $g \times g$ matrix with $\operatorname{Re} \tau < 0$, a point in the Siegel (left) half-plane $\mathcal{H}_g$. Any point $e \in \mathbb{C}^g$ can be written uniquely as $e = (\varepsilon,\delta)\begin{pmatrix} 2\pi i I \\ \tau \end{pmatrix}$ where $\delta,\varepsilon \in \mathbb{R}^g$ are the characteristics of e; the notation $e = \left\{\begin{matrix}\delta\\ \varepsilon\end{matrix}\right\}_\tau$ will be used for the point $e \in \mathbb{C}^g$ with characteristics $[e] = \begin{bmatrix}\delta\\ \varepsilon\end{bmatrix}$. If Riemann's theta function is defined by

$$\theta(z) = \theta\begin{bmatrix}0\\0\end{bmatrix}(z) = \sum_{m\in\mathbb{Z}^g} \exp\{\tfrac{1}{2}m\tau m^t + mz^t\}, \quad z \in \mathbb{C}^g \quad *$$

then for any $e = \left\{\begin{matrix}\delta\\ \varepsilon\end{matrix}\right\}_\tau \in \mathbb{C}^g$,

$$(1) \qquad \exp\{\tfrac{1}{2}\delta\tau\delta^t + (z + 2\pi i\varepsilon)\delta^t\}\theta(z+e) = \theta_\tau\begin{bmatrix}\delta\\ \varepsilon\end{bmatrix}(z)$$

$$= \sum_{m\in\mathbb{Z}^g} \exp\{\tfrac{1}{2}(m+\delta)\tau(m+\delta)^t + (z + 2\pi i\varepsilon)(m+\delta)^t\}$$

where $\theta_\tau\begin{bmatrix}\delta\\ \varepsilon\end{bmatrix}(z)$ is called the first order theta-function with characteristics $\begin{bmatrix}\delta\\ \varepsilon\end{bmatrix}$ for $\delta,\varepsilon \in \mathbb{R}^g$. In general, an nth order theta-function $\Theta_n\begin{bmatrix}\delta\\ \varepsilon\end{bmatrix}(z)$ with characteristics $\begin{bmatrix}\delta\\ \varepsilon\end{bmatrix}$ is any holomorphic function on $\mathbb{C}^g$ satisfying the identity

---

* $\exp(W) = (2.718..)^W$

$$\Theta_n\begin{bmatrix}\delta\\ \varepsilon\end{bmatrix}(z+\kappa\tau+2\pi i\lambda) = \exp\{-\tfrac{1}{2}n\kappa\tau\kappa^t - nz\kappa^t + 2\pi i(\delta\lambda^t - \varepsilon\kappa^t)\}\,\Theta_n\begin{bmatrix}\delta\\ \varepsilon\end{bmatrix}(z)$$

for $\kappa,\lambda \in \mathbb{Z}^g$; that is, for $j = 1,\ldots,g$

$$(2)\qquad \Theta_n\begin{bmatrix}\delta\\ \varepsilon\end{bmatrix}(z_1,\ldots,z_j+2\pi i,\ldots,z_g) = e^{2\pi i\delta_j}\,\Theta_n\begin{bmatrix}\delta\\ \varepsilon\end{bmatrix}(z)$$

and

$$(2)'\qquad \Theta_n\begin{bmatrix}\delta\\ \varepsilon\end{bmatrix}(z_1+\tau_{j1},\ldots,z_g+\tau_{jg}) = e^{-\frac{1}{2}n\tau_{jj} - nz_j - 2\pi i\varepsilon_j}\,\Theta_n\begin{bmatrix}\delta\\ \varepsilon\end{bmatrix}(z).$$

Any function $\Theta_n\begin{bmatrix}\delta\\ \varepsilon\end{bmatrix}$ can be considered a section of $L_\Theta^{\otimes n}$ translated by the point $\frac{1}{n}\begin{Bmatrix}\delta\\ \varepsilon\end{Bmatrix}_\tau$, where L_Θ is the line bundle on $\mathcal{A}$ defined by the divisor of Riemann's theta function [33]; for each characteristic $\begin{bmatrix}\delta\\ \varepsilon\end{bmatrix}$ there are n^g linearly independent functions $\Theta_n\begin{bmatrix}\delta\\ \varepsilon\end{bmatrix}$, and a basis is given by the functions $\theta_{n\tau}\begin{bmatrix}\frac{\delta+\rho}{n}\\ \varepsilon\end{bmatrix}(nz)$ with $\rho \in (\mathbb{Z}/n\mathbb{Z})^g$ [19, p. 40]. It is easily verified that for $\rho,\sigma \in \mathbb{Z}^g$,

$$\theta_{n\tau}\begin{bmatrix}\frac{\delta+\rho}{n}\\ \varepsilon\end{bmatrix}(nz),\qquad \theta_{\frac{\tau}{n}}\begin{bmatrix}\delta\\ \frac{\varepsilon+\sigma}{n}\end{bmatrix}(z)\qquad\text{and}\qquad \theta_\tau^n\begin{bmatrix}\frac{\delta+\rho}{n}\\ \frac{\varepsilon+\sigma}{n}\end{bmatrix}(z)$$

are all n^{th} order theta-functions with characteristics $\begin{bmatrix}\delta\\ \varepsilon\end{bmatrix}$ [19, p. 39]. In particular, when $n = 2$, $\theta_\tau^2\begin{bmatrix}\alpha\\ \beta\end{bmatrix}(z)$ is a second-order theta-function with characteristics $\begin{bmatrix}0\\ 0\end{bmatrix}$ for $2\alpha,2\beta \in \mathbb{Z}^g$; such a point $\begin{Bmatrix}\alpha\\ \beta\end{Bmatrix}_\tau \in \mathbb{C}^g$ is called a half-period and is said to be even (resp. odd) iff $\theta\begin{bmatrix}\alpha\\ \beta\end{bmatrix}$ is an even (resp. odd) function of z which holds iff $4\alpha\cdot\beta \equiv 0$ (resp. 1) mod 2. The second order theta-functions satisfy an addition theorem [17, p. 139]:

$$(3) \qquad \theta_{2\tau}\begin{bmatrix}\alpha\\ \gamma\end{bmatrix}(2z_1)\,\theta_{2\tau}\begin{bmatrix}\beta\\ \gamma\end{bmatrix}(2z_2) = \frac{1}{2^g}\sum_{2\epsilon\in(\mathbb{Z}/2\mathbb{Z})^g}(-1)^{4\alpha\cdot\epsilon}\,\theta_\tau\begin{bmatrix}\alpha+\beta\\ \gamma+\epsilon\end{bmatrix}(z_1+z_2)\,\theta_\tau\begin{bmatrix}\alpha+\beta\\ \epsilon\end{bmatrix}(z_1-z_2)$$

and its inversion

$$(4) \qquad \theta_\tau\begin{bmatrix}\alpha\\ \beta+\gamma\end{bmatrix}(z_1+z_2)\,\theta_\tau\begin{bmatrix}\alpha\\ \beta\end{bmatrix}(z_1-z_2) = \sum_{2\delta\in(\mathbb{Z}/2\mathbb{Z})^g}(-1)^{4\beta\cdot\delta}\,\theta_{2\tau}\begin{bmatrix}\delta\\ \gamma\end{bmatrix}(2z_1)\,\theta_{2\tau}\begin{bmatrix}\alpha+\delta\\ \gamma\end{bmatrix}(2z_2)$$

for any $z_1, z_2 \in \mathbb{C}^g$ and half-integer characteristics $\alpha, \beta, \gamma, \delta, \epsilon$.

Now suppose C is a compact Riemann surface of genus $g > 0$. Fix a canonical basis $A_1, \ldots, A_g, B_1, \ldots, B_g$ of $H_1(C, \mathbb{Z})$ such that the intersection matrix defined by the cup product on $H^1(C, \mathbb{Z})$ has the form $\begin{pmatrix} 0 & -I \\ I & 0 \end{pmatrix}$, and let $v_1, \ldots, v_g$ be a basis of the holomorphic differentials $H^0(C, \Omega^1_C)$ normalized so that the period matrix of $v_1, \ldots, v_g$ with respect to $A_1, \ldots, A_g, B_1, \ldots, B_g$ has the form $(2\pi i I, \tau)$ where I = identity $g \times g$ matrix and τ is a symmetric $g \times g$ matrix with $\mathrm{Re}\,\tau < 0$. Then $J(C) = \mathbb{C}^g/(2\pi i I, \tau)$, the Jacobian variety of C, is identified with the group of divisors on C of degree 0 modulo principal divisors: the divisor $D = \mathcal{B} - \mathcal{A}$ with $\mathcal{A}$ and $\mathcal{B}$ positive divisors of the same degree corresponds to the point $\left(\int_{\mathcal{A}}^{\mathcal{B}} v\right)$ in $J(C)$. Equivalently, $J(C)$ will often be identified with the group of holomorphic equivalence classes of line bundles on C of degree 0 as follows: if $\chi: \pi_1(C) \to \mathbb{C}^{*2g}$ is the characteristic homomorphism defining a flat line bundle [13, p. 186], the bundle of degree 0 corresponding to $D = \mathcal{B} - \mathcal{A} \in J(C)$ will have characteristic homomorphism

$$(5) \qquad \chi(A_i) = 1 \qquad \chi(B_i) = \exp\left(\int_{\mathcal{A}}^{\mathcal{B}} v_i\right) \qquad i = 1, \ldots, g;$$

two different paths of integration from $\mathcal{A}$ to $\mathcal{B}$ give rise to holomorphically equivalent bundles, since two flat line bundles L and $\tilde{L}$

are holomorphically equivalent if and only if, for some $\omega \in H^0(\Omega_C^1)$, their characteristic homomorphisms χ and $\tilde{\chi}$ satisfy $\tilde{\chi}(\gamma)\chi^{-1}(\gamma) = \exp\int_\gamma \omega$, $\forall\ \gamma \in \pi_1(C)$ [14, p. 238]. This last fact also implies that the line bundle L is holomorphically equivalent to the unitary line bundle with characteristic homomorphism

$$(6) \qquad \tilde{\chi}(A_j) = e^{-2\pi i\delta_j}, \qquad \tilde{\chi}(B_j) = e^{2\pi i\varepsilon_j}, \qquad j = 1,\dots,g$$

if $\begin{bmatrix}\delta\\ \varepsilon\end{bmatrix}_\tau$ are the characteristics of the point $(\int_{\mathcal{A}}^{\mathcal{B}} v) \in \mathbb{C}^g$.

For any $a \in C$ and $e \in \mathbb{C}^g$, the multiplicative meromorphic functions $\theta(\int_a^x v - \int_{\mathcal{A}}^{\mathcal{B}} v - e)/\theta(\int_a^x v - e)$ and $\theta[-\int_{\mathcal{A}}^{\mathcal{B}} v](\int_a^x v - e)/\theta(\int_a^x v - e)$ are, by (2), meromorphic sections of L as given by the characteristic homomorphisms (5) and (6) respectively. Alternatively, meromorphic sections of L can be expressed in terms of the Abelian integrals as follows: if $D = \sum n_i p_i$, $p_i \in C$, is any divisor of degree 0 and ω_D is the unique differential of the third kind on C with simple poles of residue n_i at p_i and zero A-periods, the Riemann bilinear relation gives

$$(7) \qquad \int_{B_j} \omega_D = \int_{\mathcal{A}}^{\mathcal{B}} v_j \in \mathbb{C} \qquad j = 1,\dots,g$$

where $\mathcal{B} = \sum_{n_i>0} n_i p_i$ and $\mathcal{A} = \sum_{n_i<0} n_i p_i$. Here, and henceforth, we make the convention that the paths of integration between divisors are taken within C cut along its homology basis, which we assume does not contain points of D. By (7), the multiplicative meromorphic function $\exp\int^x \omega_D$ is thus a meromorphic section of the line bundle corresponding to D under (5). Similarly, if

$$\Omega_D(x) = \omega_D(x) - \sum_{i,j=1}^{g} v_i(x)(\operatorname{Re}\tau)^{-1}_{ij} \operatorname{Re}\,(\int_{\mathcal{A}}^{\mathcal{B}} v_j)$$

is the unique differential of the third kind with simple poles of residue n_i at p_i and purely imaginary A and B-periods, then $\exp \int^x \Omega_D$ is a meromorphic section of the line bundle with characteristic homomorphism (6). Finally, whenever the bundle is holomorphically trivial, we have Abel's Theorem: if $D = \mathcal{B} - \mathcal{A}$ is the divisor of a meromorphic function f on C,

$$d \ln f = \omega_{\mathcal{B}-\mathcal{A}} - \sum_1^g m_j v_j$$

$$\int_{\mathcal{A}}^{\mathcal{B}} v_k = 2\pi i n_k + \sum_{j=1}^g m_j \tau_{jk} \tag{8}$$

$$m_j = -\frac{1}{2\pi i}\int_{A_j} d \ln f \in \mathbb{Z}, \qquad n_j = \frac{1}{2\pi i}\int_{B_j} d \ln f \in \mathbb{Z}.$$

For $n \in \mathbb{Z}$, let $J_n(C)$ be the principal homogeneous space of linear equivalence classes of divisors of degree n on C. Then $J(C) = J_0(C)$ acts on $J_n(C)$ by addition, and for any fixed divisor $\mathcal{B} \in J_n(C)$ there is an isomorphism $\psi_{\mathcal{B}} : J_n(C) \simeq J_0(C)$ given by $\psi_{\mathcal{B}}(\mathcal{A}) = \mathcal{A} - \mathcal{B}$ for $\mathcal{A} \in J_n(C)$. For two divisors $\mathcal{A}$ and $\mathcal{B}$ of the same degree n, we will write $\mathcal{A} = \mathcal{B}$ whenever $\mathcal{A}$ and $\mathcal{B}$ are the same point in $J_n(C)$ - that is, $\mathcal{A} - \mathcal{B}$ is the divisor of a meromorphic function on C. Frequently $J_n(C)$ will be identified with the space of (holomorphic) equivalence classes of holomorphic line bundles on C of degree n as follows: to any divisor D of degree n and open covering $\{U_i\}$ of C, associate the line bundle of degree n given by the cocycle $\left(\frac{d_i}{d_j}\right) \in H^1(C, \mathcal{O}_C^*)$, where d_i, d_j are meromorphic functions on U_i, U_j with divisor D there. If $\mathcal{D}$ is the sheaf of germs of meromorphic functions on C with divisors at least -D, there is an isomorphism $H^0(D) \simeq H^0(\mathcal{D})$ by sending $(s_i) \in H^0(D)$ to $\left(\frac{s_i}{d_i}\right) \in H^0(\mathcal{D})$. Alternatively, holomorphic sections of D will often be considered holomorphic functions on the universal cover $U \xrightarrow{\pi} C$ by lifting $(s_i) \in H^0(D)$ to $\left(\frac{\pi^* s_i}{\sigma_i}\right) \in H^0(U, \mathcal{O}_U)$ for some trivialization $(\sigma_i) \in H^0(U, \pi^*(D))$ of the induced bundle $\pi^*(D)$ on U.

If $P_n \subseteq J_n(C)$ is the set of all classes of positive divisors of degree $n > 0$ on C, and $C^{(n)}$ is the symmetric product of C with itself n times, the mapping

$$\sigma_n: (x_1, \ldots, x_n) \in C^{(n)} \longrightarrow \sum_1^n x_i \in P_n$$

onto P_n is $1 - 1$ at those points $D = \sum_1^n x_i$ for which the index of speciality $i(D) = g - n$; and, in general, it can be shown [22] that the rank of σ_n at $(x_1, \ldots, x_n) \in C^{(n)}$ is equal to $g - i(\sum_1^n x_i)$. The Jacobi inversion theorem says that σ_g is onto $J_g(C)$ which means, from the isomorphism $J_g \simeq J_0$, that for any point $e \in J_0(C)$ and divisor $\mathcal{A}$ of degree g, there is a positive divisor $\mathcal{B}$ of degree g with $e = \mathcal{B} - \mathcal{A}$ in J_0. The map σ_g fails to be $1 - 1$ for those divisors D of degree g with $i(D) > 0$; by the Riemann-Roch theorem, this is the $g - 2$ dimensional subvariety of J_g described by $K_C - D'$, where D' is any positive divisor of degree $g - 2$ and $K_C \in J_{2g-2}(C)$ is the class of the divisor of any differential on C [32, p. 73]. The following fundamental theorem of Riemann says that $P_{g-1} \subset J_{g-1}(C)$ is a translate of the divisor of the classical theta-function $(\theta) \subset J_0(C)$ by a fixed class $\Delta \in J_{g-1}(C)$ and that the degree of the tangent cone of P_{g-1} at a point $D \in P_{g-1}$ is $\operatorname{mult}_{D-\Delta}(\theta) = i(D)$; in particular, $\operatorname{div}_C \theta(\mathcal{A} - x - \Delta) = \mathcal{A}$ for any positive divisor $\mathcal{A}$ of degree g on C with $i(\mathcal{A}) = 0$, while $\theta(\mathcal{A} - x - \Delta) \equiv 0$ on C if $\mathcal{A}$ is any divisor of degree g with $i(\mathcal{A}) > 0$. For a proof of this important theorem, see either the classical presentation in Lewittes [21], or Mayer's thesis [22].

<u>Theorem 1.1</u> (Riemann). There is a divisor class $\Delta \in J_{g-1}(C)$ with

$$2\Delta = K_C \in J_{2g-2}(C) \tag{9}$$

such that for any $a \in C$ and $e \in \mathbb{C}^g$:

i) If $\Theta(e) \neq 0$, then $\operatorname{div}_C \Theta(x-a-e) = \mathcal{A}$ is positive of degree g with $i(\mathcal{A}) = 0$ and

$$e = \mathcal{A} - a - \Delta \qquad \text{in } J_0(C). \tag{10}$$

ii) If $\Theta(e) = 0$, then for some positive divisor ζ of degree $g-1$,

$$e = \zeta - \Delta \qquad \text{in } J_0(C). \tag{11}$$

Here $i(\zeta)$ is the multiplicity of (Θ) at e and is the smallest integer d such that $\Theta(\mathcal{B} - \mathcal{A} - e) \equiv 0$ for all positive divisors $\mathcal{A}, \mathcal{B}$ of degree $\leq$ d-1. In case $i(\zeta) = 1$ and $i(\zeta+a) = 0$, $\operatorname{div}_C \Theta(x-a-e) = a+\zeta$; in all other cases, $\Theta(x-a-e) \equiv 0$ on C.

Generally, Δ is not the class of a positive divisor on C; otherwise (11) would imply that $\Theta(0) = 0$, which is possible only for curves of special moduli since the variety $\Theta(0) = 0$ in $\mathfrak{H}_g$ does not contain, say, period matrices of curves of genus g which are nearly products of g elliptic curves (see Cor. 3.2). However, for any $a \in C$, Δ is the class of the divisor of a half-order differential with only a simple pole at a, called the Szego-kernel of C with characteristic $\begin{bmatrix} 0 \\ 0 \end{bmatrix}$ (Cor. 2.12). Although Δ depends on the choice of canonical homology basis defining Θ, if the homology is changed to $\begin{pmatrix} \tilde{A} \\ \tilde{B} \end{pmatrix} = \begin{pmatrix} d & c \\ b & a \end{pmatrix} \begin{pmatrix} A \\ B \end{pmatrix}$ with $\begin{pmatrix} a & b \\ c & d \end{pmatrix} \in Sp(2g, \mathbb{Z})$, the transformation law of theta functions [17, p. 84] says that

$$\Theta_{\tilde{T}}\begin{bmatrix} \tilde{g} \\ \tilde{h} \end{bmatrix}(\tilde{z}) = \kappa (\det M)^{1/2} \exp\left\{ \frac{1}{2} \sum_{i \leq j} z_i z_j \frac{\partial \ln \det M}{\partial T_{ij}} \right\} \Theta_T \begin{bmatrix} g \\ h \end{bmatrix}(z) \tag{12}$$

where $M = (c\tau + 2\pi i d)$, $2\pi i z = \tilde{z}M \in \mathbb{C}^g$, $\tilde{\tau} = 2\pi i(a\tau + 2\pi i b)(c\tau + 2\pi i d)^{-1}$, κ is some constant independent of $z \in \mathbb{C}^g$ and $\tau \in \mathcal{H}_g$ and

$$\begin{pmatrix} \tilde{g} \\ \tilde{h} \end{pmatrix} = \begin{pmatrix} d & -c \\ -b & a \end{pmatrix}\begin{pmatrix} g \\ h \end{pmatrix} + \tfrac{1}{2}\,\mathrm{Diag}\begin{pmatrix} cd^t \\ ab^t \end{pmatrix} \in \mathbb{R}^{2g},$$

"Diag" denoting the column vector of the diagonal entries to $\begin{pmatrix} cd^t \\ ab^t \end{pmatrix}$. As a result Δ is transformed to $\tilde{\Delta} \in J_{g-1}(C)$ where $\tilde{\Delta} - \Delta$ is the (necessarily even) half-period in $J_0(C) = \mathbb{C}^g/(2\pi i I, \tau)$ given by $\left\{\tfrac{1}{2}\,\mathrm{Diag}\begin{pmatrix} cd^t \\ ab^t \end{pmatrix}\right\}_\tau$.

Classically, Δ was constructed as follows: if genus $C > 1$ and $C = U/\Gamma$ with $U = \{z \in \mathbb{C} \mid |z| < 1\}$ and Γ a Fuchsian group of the first kind with compact fundamental domain, then the cycles A_i on C can be lifted to paths from z_0 to $g_{A_i}(z_0)$ in U with $z_0 \in U$ some fixed base point and $g_{A_i} \in \Gamma$. Then for any $a \in C$ and $j = 1,2,\ldots,g$, the sum $\sum_{\substack{i=1 \\ i\neq j}}^{g} \int_{z_0}^{g_{A_i}(z_0)} (v_i(x) \int_a^x v_j)$ is independent of the basepoint z_0 so that we can unambiguously define the vector $k^a = (k_1^a,\ldots,k_g^a) \in \mathbb{C}^g$ of Riemann constants with basepoint $a \in C$ by

$$k_j^a = \frac{2\pi i - \tau_{jj}}{2} + \frac{1}{2\pi i}\sum_{\substack{i=1 \\ i\neq j}}^{g} \int_{A_j} (v_i(x) \int_a^x v_j) \tag{13}$$

Since $k^b - k^a = (g-1)\int_b^a v \in \mathbb{C}^g$ for any two points $a,b \in C$, the divisor class $k^a + (g-1)a$ is independent of $a \in C$; it is a consequence of the classical theory that:

$$\Delta = (g-1)a + k^a \in J_{g-1}(C). \tag{14}$$

More precisely, the relation in $\mathbb{C}^g$, giving (10) under the projection

$\mathbb{C}^g \to J_0(C)$, is a special case of the following version of Riemann's Theorem (10) for differentials of the second kind, proved by applying Cauchy's residue theorem on the simply connected surface C^0 obtained by fixing a basepoint $p_0 \in C$ and dissecting C along loops in $\pi_1(C,p_0)$ homologous to the cycles $A_1,B_1,\ldots,A_g,B_g$ [14, p. 151].

Proposition 1.2. Let $e \in \mathbb{C}^g$ and suppose $\mathrm{div}_C\Theta(x-a-e) = \mathcal{A} \neq C$ for some point $a \in C$. Then for any $p \in C$ and any differential ω of the second kind on C with at most poles at $b_1,\ldots,b_N \in C$,

$$\begin{aligned} \frac{1}{2\pi i}\sum_{k=1}^{g} e_k \int_{A_k}\omega &= \int_{gp}^{\mathcal{A}}\omega + \sum_{j=1}^{N} \operatorname*{Res}_{x=b_j}\left(\int^x \omega \cdot d\ln\Theta\left(\int_a^x v - e\right)\right) \\ &+ \frac{1}{2\pi i}\sum_{k=1}^{g}\left[\left(\frac{\tau_{kk}}{2} + \int_a^{p_0} v_k\right)\int_{A_k}\omega - \int_{A_k} v_k(y)\int_p^y \omega\right] \\ &+ \sum_{k=1}^{g}\left[(m_k-1)\int_{B_k}\omega - (n_k-1)\int_{A_k}\omega\right] \quad \text{in } \mathbb{C} \end{aligned} \tag{15}$$

where

$$\begin{aligned} m_k &= \frac{1}{2\pi i}\int_{A_k} d\ln\Theta\left(\int_a^x v - e\right) \in \mathbb{Z} \\ n_k &= \frac{1}{2\pi i}\left[\frac{\tau_{kk}}{2} + \int_a^{p_0} v_k - e_k + \int_{B_k} d\ln\Theta\left(\int_a^x v - e\right)\right] \in \mathbb{Z}, \end{aligned} \tag{16}$$

and where all integrals are taken within C dissected along $\pi_1(C,p_0)$, which we assume does not contain $\mathcal{A}$ or the points $a,p,b_1,\ldots,b_N$. In particular, when ω is the normalized holomorphic differential v_j, $j = 1,\ldots,g$, and when $m_k,n_k \in \mathbb{Z}$ are defined by (16),

$$e_j = \int_{ga}^{\mathcal{A}} v_j - \frac{1}{2\pi i}\sum_{\substack{k=1\\k\neq j}}^{g}\int_{A_k}\left(v_k\int_a^x v_j\right) + \frac{\tau_{jj}-2\pi i}{2} + \sum_{k=1}^{g}(m_k-1)\tau_{kj} - 2\pi i(n_j-1)$$

which gives (14) and the relation (10) in $J_0(C)$.

From the fundamental property (9) of Δ, we have

Corollary 1.3. If $a,b \in C$ and $e \in \mathbb{C}^g$ with $\Theta(e) \neq 0$, then $\Theta(x-a+e)\Theta(x-b-e)$ vanishes at the 2g zeroes of a differential of the third kind on C with poles at a and b if $a \neq b$, and at the 2g zeroes of a differential of the second kind with a double pole at a if $a = b$. This latter differential will have g double zeroes if e is a half-period.

Corollary 1.4. If $f \in \mathbb{C}^g$ with $\Theta(f) = 0$ then $\Theta(x-a-f)\Theta(x-b+f)$ either vanishes identically on C or vanishes at $x = a$ and $x = b$ and at the $2g-2$ zeroes of the holomorphic differential $H_f(x) = \sum_{i=1}^{g} \frac{\partial\Theta}{\partial z_i}(f)v_i(x)$, independent of $a,b \in C$. The latter situation holds, by Riemann's Theorem (ii), if and only if $f \in (\Theta)$ is a non-singular point and $i(\Delta+a+f) = i(\Delta+b-f) = 0$.

Proof. Let $f \in (\Theta)$ be non-singular and assume that $\Theta(x-a-f)\Theta(x-b+f) \not\equiv 0$ on C; then $\operatorname{div}_C\Theta(x-a-f) = a+\zeta$ and $\operatorname{div}_C\Theta(x-b+f) = b+\xi$ with ξ,ζ positive divisors of degree $g-1$. From (11), $f = \zeta-\Delta$ and $-f = \xi-\Delta$ so that ξ and ζ are independent of the points a, b since $i(\zeta) = i(\xi) = 1$. Furthermore, $\xi+\zeta = 2\Delta = K_C$ so $\operatorname{div}_C\Theta(x-a-f)\Theta(x-b+f) = a + b + \operatorname{div}_C H_f(x)$ for some holomorphic differential H_f independent of a, b. To find $H_f(x)$, specialize to $a = b$; then by symmetry in x and a, $\frac{\Theta(x-a-f)\Theta(x-a+f)}{H_f(x)H_f(a)}$ is, in the variables x and a, a holomorphic section of a line bundle on $C \times C$ with a double zero along the diagonal $x = a$ and with no other zeroes. By taking a Taylor expansion of this section near the point $x = a$, we conclude that $\frac{1}{H_f(a)} \sum_1^g \frac{\partial\Theta}{\partial z_i}(f)v_i(a)$ is holomorphic and non-zero $\forall\, a \in C$ so that $H_f(x)$ is, up to a constant, $\sum_1^g \frac{\partial\Theta}{\partial z_i}(f)v_i(x)$.

If we define the open $g-1$ dimensional subvariety

$$\Theta_{\text{non-sing}} = \{\zeta-\Delta \in (\Theta) \mid \zeta \text{ positive of degree } g-1 \text{ with } i(\zeta) = 1\},$$

then the map $\Theta_{\text{non-sing}} \to \mathbb{P}(H^0(\Omega_C^1)) \simeq \mathbb{P}_{g-1}(\mathbb{C})$ sending $f \to H_f = \sum_1^g \frac{\partial\theta}{\partial z_i}(f)v_i$ is of degree $\binom{2g-2}{g-1}$ and is ramified at the points $\{\zeta - \Delta \in (\Theta) \mid \zeta$ is positive of degree $g-1$ with at least two points in ζ the same$\}$; the image of this set in $\mathbb{P}(H^0(\Omega_C))$ consists of differentials H_f with at least one double zero, which is the dual curve to the canonical imbedding of a non-hyperelliptic C in $\mathbb{P}_{g-1}(\mathbb{C})$ [4, p. 820]. Andreotti and Mayer [5] have shown that $\dim \Theta_{\text{sing}}$, the singular set of (Θ), is $g-4$ for curves of genus ≥ 4 except in the hyperelliptic case, where it has dimension $g-3$ (p. 13 below); the set of all points $f = \sum_1^{g-1} p_i - \Delta \in (\Theta)$ with the $p_i \in C$ distinct and $i(\sum_1^{g-1} p_i) = 2$ is dense in Θ_{sing} and $\sum_{i,j=1}^{g} \frac{\partial^2\theta}{\partial z_i \partial z_j}(f)z_i z_j = 0$ is the equation of a quadric surface of rank ≤ 4 containing the canonical curve in $\mathbb{P}_{g-1}(\mathbb{C})$. (See Cor. 2.18.)

The group of 4^g half-periods in $J_0(C)$ acts effectively on the set of 4^g holomorphically inequivalent line bundles $L \in J_{g-1}(C)$ with $L \otimes L = K_C$. We will denote by L_0 the bundle corresponding to the divisor class $\Delta \in J_{g-1}(C)$ on p. 7 and let L_α be that bundle for which $L_\alpha \otimes L_0^{-1}$ is the half-period $\alpha \in J_0(C)$. Then Riemann's Theorem 1.1 for half-periods implies

Corollary 1.5.* $\dim H^0(L_\alpha) = \dim H^1(L_\alpha) = \text{mult}_\alpha \theta$ is even for α even and odd for α odd.

For generic Riemann surfaces, $\dim H^0(L_\alpha)$ is 0 for the $\frac{1}{2}(4^g + 2^g)$ even α and 1 for the $\frac{1}{2}(4^g - 2^g)$ odd α. By Cor. 1.4, the holomorphic section of L_α for α odd and non-singular is, up to a constant, given

---

* For a discussion of this corollary in terms of spin-structures, see [6].

by the half-order differential $h_\alpha \in H^0(L_\alpha)$ satisfying $h_\alpha^2(x) = \sum_1^g \frac{\partial\Theta[\alpha]}{\partial z_i}(0)v_i(x)$. A meromorphic section of L_β for β even and non-singular, with a simple pole at $x = a \in C$, will then be given by $\frac{\Theta[\beta](x-a)}{\Theta[\alpha](x-a)} h_\alpha(x)$. In general, the number of linearly independent holomorphic Abelian ("root") functions of degree n is, by the Riemann-Roch formula,

$$\dim H^0(\prod_1^n L_{\alpha_i}) = (n-1)(g-1) + \begin{cases} 1 \text{ if } n=1,\ \alpha_1 \text{ odd; or } n=2,\ \alpha_1=\alpha_2 \\ 0 \text{ otherwise.} \end{cases}$$

For example, when $g = 3$, $n = 2$ and $\alpha_1 \neq \alpha_2$, any linear relation among three Prym differentials (sections of $K_C \otimes (\alpha_1 + \alpha_2)$) gives the quartic equation for non-hyperelliptic C as a sum of three square roots of quadratic polynomials in the homogeneous coordinates of $\mathbb{P}_2(\mathbb{C})$ - see p. 80 and [7, p. 387]. In terms of a plane model of C of genus g and degree n, the g-1 points $\operatorname{div}_C h_\alpha$ are the points of tangency of an adjoint curve of degree n-3, while the g points $\operatorname{div}_C \Theta[\beta](x-a)$ are the points of tangency of an adjoint of degree n-2 through the points of intersection of C with a line tangent to C at a. An algorithm for computing these adjoints from the coefficients of the equation for C is given in [30].

<u>Example</u>. Let C be the hyperelliptic curve $s^2 = \prod_1^{2g+2} (z - z(Q_i))$ given as a two-sheeted covering of the Riemann sphere by the function z with ramification at the Weierstrass points $Q_1,\ldots,Q_{2g+2} \in C$. If ϕ is the involution on C interchanging the two sheets, the divisor class $D = x + \phi(x) \in J_2(C)$ is independent of $x \in C$ and satisfies

$$(g+1)D = \sum_1^{2g+2} Q_i \quad \text{and} \quad (g-1)D = K_C = 2\Delta.$$

For any $x_1,\ldots,x_{g+1} \in C$, $\Theta(\sum_1^{g+1} x_i - D - \Delta)$ vanishes exactly when

$x_i = \phi(x_j)$ for some $i \neq j$, and by Riemann's Theorem 1.1 and the Weierstrass gap theorem [21, p. 60], the order $i(\sum_1^{g+1} x_i - D)$ to which Θ vanishes at such a point is the total number of disjoint pairs of points (x_i, x_j) for which $x_i = \phi(x_j)$, $i \neq j$. Thus Θ has multiplicity exactly m on the subvariety of J_0 given by all points $\sum_1^{g+1-2m} x_i + (m-1)D - \Delta$ with $x_i \in C$, $x_i \neq \phi(x_j)$ for $i \neq j$; in particular, the 4^g half-periods $\begin{bmatrix}\mu\\ \nu\end{bmatrix}$, $2\mu, 2\nu \in \mathbb{Z}^g$, are in 1-1 correspondence with the 4^g partitions $\{i_1,\ldots,i_{g+1-2m}\} \cup \{j_1,\ldots,j_{g+1+2m}\}$ of $\{1,\ldots,2g+2\}$ for integers $m \geq 0$ as follows: for $m = 0$, $\sum_{m=1}^{g+1} Q_{i_k} - D - \Delta$ give the $\frac{1}{2}\binom{2g+2}{g+1} = \binom{2g+1}{g}$ non-singular even half-periods; while for $m = 1$, $\sum_{k=1}^{g-1} Q_{i_k} - \Delta$ are the $\binom{2g+2}{g-1}$ non-singular odd half-periods and for $m > 1$, $\sum_{k=1}^{g+1-2m} Q_{i_k} + (m-1)D - \Delta$ are the $\binom{2g+2}{g+1-2m}$ singular half-periods of multiplicity m which are even (odd) for m even (odd). In the case of the non-singular half-periods, the corresponding half-order differentials mentioned on p. 12 can be given explicitly: if $\alpha = \sum_1^{g-1} Q_{i_k} - \Delta$ is an odd half-period then $h_\alpha^2(x) = \sum_1^g \frac{\partial\Theta[\alpha]}{\partial z_i}(0)v_i(x)$ is, up to a constant, given by the differential $\prod_1^{g-1} (z(x) - z(Q_{i_k})) \frac{dz(x)}{s(x)}$ with g-1 double zeroes at $Q_{i_1},\ldots,Q_{i_{g-1}}$. On the other hand, if $\beta = \sum_1^{g+1} Q_{i_k} - D - \Delta$ is an even half-period and $\psi(x) = \prod_{k=1}^{g+1} \frac{z(x) - z(Q_{i_k})}{z(x) - z(Q_{j_k})}$, then

$$(17) \qquad m_\beta(x,y) = \frac{1}{2}\left(\sqrt[4]{\frac{\psi(x)}{\psi(y)}} + \sqrt[4]{\frac{\psi(y)}{\psi(x)}}\right)\frac{\sqrt{dz(x)\,dz(y)}}{z(y)-z(x)}$$

$$= \frac{s(y)\prod_1^{g+1}(z(x)-z(Q_{i_k})) + s(x)\prod_1^{g+1}(z(y)-z(Q_{i_k}))}{2(z(y)-z(x))} \cdot \left[\frac{dz(x)\,dz(y)}{s(x)\,s(y)\prod_1^{g+1}(z(x)-z(Q_{i_k}))(z(y)-z(Q_{i_k}))}\right]^{1/2}$$

is a bilinear half-order differential on $C \times C$, antisymmetric in x and y and with a simple pole only along $y = x$; specializing $x = Q_{i_1}$ say, $\mathrm{div}_C\, m_\beta(Q_{i_1}, y) = \sum_1^{g+1} Q_{i_k} - 2Q_{i_1} = \beta + \Delta$, so that $\Theta[\beta]\left(\int_x^y v\right)$ vanishes at the g zeroes of $m_\beta(x,y)$, the Szego kernel for the half-period β. If we set $E(x,y) = \Theta[\beta]\left(\int_x^y v\right)/\Theta[\beta](0)m_\beta(x,y)$, then $E(x,y)$ is independent of the non-singular even half-period β and is holomorphic with only a zero of first order along $y = x$; from $E(x,y)$ we can in turn construct the normalized differentials of the second and third kind on C - see §2.

It is convenient to have the explicit value of the Riemann constants k^a for some $a \in C$: for simplicity, take a to be the Weierstrass point Q_1 and choose a symmetric canonical homology basis $A_1, B_1, \ldots, A_g, B_g$ as in the picture with $\phi(A_i) = -A_i$ and $\phi(B_j) = -B_j$ in $H_1(C, \mathbb{Z})$. The corresponding normalized differentials $v_1, \ldots, v_g$ satisfy $\phi^* v_i = -v_i$, $i = 1, \ldots, g$ so that

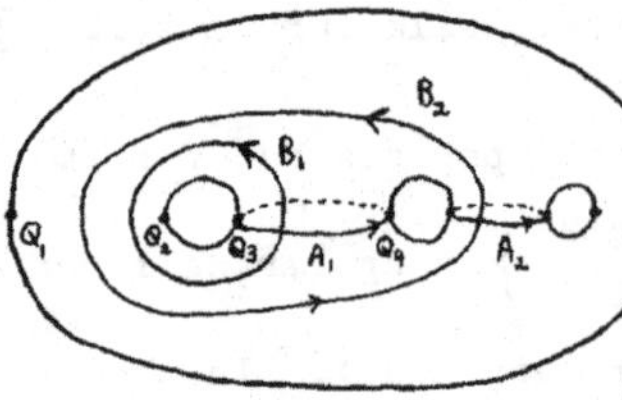

g=3

$$\sum_{k=1}^{g} \left\{ \int_{Q_{2k+1}}^{Q_{2k+2}} \left(v_k(x) \int_{Q_{2k+1}}^{x} v_j \right) + \int_{Q_{2k+2}}^{Q_{2k+1}} \left(v_k(\phi(x)) \int_{Q_{2k+1}}^{\phi(x)} v_j \right) \right\} = 0 \qquad j = 1, \ldots, g$$

The Riemann constants for the basepoint Q_1 are therefore

$$k_j^{Q_1} = \frac{-T_{jj} + 2\pi i}{2} + \frac{1}{2\pi i} \sum_{\substack{k=1 \\ k \neq j}}^{g} \int_{A_k} \left(v_k \int_{Q_1}^{Q_{2k+1}} v_j \right)$$

(18)

$$= -\frac{1}{2} T_{jj} + (2-g)\pi i + \sum_{\substack{k=1 \\ k \neq j}}^{g} \int_{Q_2}^{Q_{2k+1}} v_j \qquad = -\frac{1}{2} \sum_{k=1}^{g} T_{jk} + \pi i (2-j)$$

and $\Delta - (g-1)Q_1 = k^{Q_1} = \begin{Bmatrix} \frac12 & \frac12 & \frac12 & \frac12 & \cdots \\ \frac12 & 0 & \frac12 & 0 & \cdots \end{Bmatrix}_\tau \in J_0(C)$, a half-period which is odd for $g \equiv 1$ or $2 \bmod 4$ and even for $g \equiv 0$ or $3 \bmod 4$. This fact, together with the relations

$$Q_1 - Q_2 = \begin{Bmatrix} 0 & 0 & \ldots & 0 \\ \tfrac{1}{2} & \tfrac{1}{2} & \ldots & \tfrac{1}{2} \end{Bmatrix}_\tau, \qquad Q_{2k+1} - Q_2 = \begin{Bmatrix} 0 & \ldots & 0 & \overset{(k)}{-\tfrac{1}{2}} & 0 & \ldots & 0 \\ \tfrac{1}{2} & \ldots & \tfrac{1}{2} & 0 & 0 & \ldots & 0 \end{Bmatrix}_\tau$$

and $$Q_{2k+2} - Q_{2k+1} = \begin{Bmatrix} 0 & \ldots & \overset{(k)}{0} & \ldots & 0 \\ 0 & \ldots & \tfrac{1}{2} & \ldots & 0 \end{Bmatrix}_\tau, \qquad 1 \leq k \leq g$$

in $J_0(C)$, determine the half-period corresponding to any partition as described above.

Further discussion of hyperelliptic Θ-functions can be found in [7, Ch. XI], which contains some addition-theorems of §2; and in the survey article by Krazer-Wirtinger [20], a general reference for the entire classical theory. An excellent account of the varieties of special divisors in $J_0(C)$, including the hyperelliptic case, can be found in the recent notes on Jacobi varieties by R. Gunning [37].

II. The Prime-Form

In order to give the relations between theta-functions and Abelian differentials on C, we need

Definition 2.1. Let α be a non-singular odd half-period corresponding to a bundle L_α as on p. 11 with a holomorphic section h_α satisfying $h_\alpha^2(x) = \sum_{i=1}^{g} \frac{\partial\Theta[\alpha]}{\partial z_i}(0)v_i(x)$. Then the prime-form is given by

$$E(x,y) = \frac{\Theta[\alpha](y-x)}{h_\alpha(x)h_\alpha(y)} \qquad \forall\ x,y \in C. \tag{19}$$

Remark. The existence of a non-singular odd half-period follows from the fact that the h_α^2 actually span $H^0(C,\Omega_C^1)$ (Cor. 4.21). An alternate definition of the prime-form, independent of Θ-functions, is provided by (v) below.

From the properties of Riemann's theta-function given in §1, it is easily seen that:

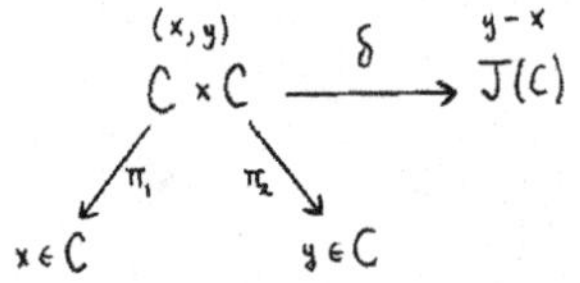

i) E(x,y) is independent of α and is a holomorphic section of the bundle $\pi_1^*L_0^{-1} \otimes \pi_2^*L_0^{-1} \otimes \delta^*(\Theta)$ on $C \times C$, where π_1 and π_2 are the projections of $C \times C$ onto its first and second factors, and δ is the map sending $(x,y) \in C \times C$ into $y-x \in J(C)$. Thus, for fixed $x \in C$, E(x,y) is a multiplicative $-\frac{1}{2}$ order differential in y with multipliers along the A_i and B_j cycles in y given by:

$$1 \text{ and } \exp\left(-\frac{\tau_{jj}}{2} - \int_x^y v_j\right) \text{ respectively.} \tag{20}$$

ii) $E(x,y) = -E(y,x)$ $\forall\ x,y \in C$ and E(x,y) vanishes to first order along the diagonal $y = x$ in $C \times C$ and is otherwise non-zero.

iii) If $\mathcal{A} = \sum_1^n a_i$ and $\mathcal{B} = \sum_1^n b_i$ with $a_i, b_i \in C$,

(21) $$d \ln \prod_1^n \frac{E(x,b_i)}{E(x,a_i)} = \omega_{\mathcal{B}-\mathcal{A}}(x) \qquad \forall\, x \in C$$

so that $\prod_1^n \frac{E(x,b_i)}{E(x,a_i)}$ is a meromorphic section of the line bundle corresponding to the divisor $\mathcal{B} - \mathcal{A}$ under (5). This gives the classical "interchange" law:

(22) $$\int_X^Y \omega_{\mathcal{B}-\mathcal{A}} = \int_{\mathcal{A}}^{\mathcal{B}} \omega_{Y-X}$$

for any divisors $\mathcal{A}$, $\mathcal{B}$, X and Y of degree n on C.

iv) Though E(x,y) depends on the choice of homology basis defining θ, a change of basis $\begin{pmatrix} \tilde{A} \\ \tilde{B} \end{pmatrix} = \begin{pmatrix} d & c \\ a & b \end{pmatrix} \begin{pmatrix} A \\ B \end{pmatrix}$, with $\begin{pmatrix} a & b \\ c & d \end{pmatrix} \in Sp(2g,\mathbb{Z})$ as in (12), transforms E(x,y) into

$$\tilde{E}(x,y) = E(x,y) \exp\left\{ \frac{1}{2} \sum_{i \le j} \frac{\partial}{\partial \tau_{ij}} \ln \det(c\tau + 2\pi i d) \int_x^y v_i \int_x^y v_j \right\}$$

v) If C is realized as a covering of the Riemann sphere by a meromorphic function $z: C \to \mathbb{P}_1(\mathbb{C})$,

$$E^2(x,y) = \frac{(z(y)-z(x))^2}{dz(x)\,dz(y)} \exp\left\{ \int_{-x+z^{-1}z(x)}^{-y+z^{-1}z(y)} \omega_{y-x} + \sum_{i=1}^{g} \int_x^y m_i v_i \right\} \qquad \forall\, x,y \in C$$

where $m_j = \frac{1}{2\pi} \int_{A_j} d \arg \frac{z - z(y)}{z - z(x)}$ and paths of integration are taken within C cut along its homology basis. This classical formula is proved as follows: for p,q near $x,y \in C$,

$$\frac{z(y)-z(q)}{z(y)-z(p)} \cdot \frac{z(x)-z(p)}{z(x)-z(q)} = \exp\left\{ \int_{-p+z^{-1}z(p)}^{-q+z^{-1}z(q)} \omega_{y-x} + \int_p^q \omega_{y-x} + \sum_{i=1}^{g} \int_x^y m_i v_i \right\} \qquad \forall\, x,y \in C$$

by (22) and Abel's Theorem (8); now let $q \to y$ and $p \to x$, using the fact that $\exp \int_p^q \omega_{y-x} = \frac{E(y,q)E(x,p)}{E(x,q)E(y,p)}$ by (21).

The holomorphic Prym differentials [14, p. 160] with g-1 double zeroes on C are given by:

Proposition 2.2. For any non-singular point $f \in (\Theta)$, let

$$H_f(x) = \sum_{i=1}^{g} \frac{\partial\Theta}{\partial z_i}(f)v_i(x) \quad \text{and} \quad Q_f(x) = \sum_{i,j=1}^{g} \frac{\partial^2\Theta}{\partial z_i \partial z_j}(f)v_i(x)v_j(x).$$

Then for all $x,y \in C$,

$$\left(\frac{\Theta(y-x-f)}{E(x,y)}\right)^2 = H_f(x)H_f(y) \exp \int_y^x \frac{Q_f}{H_f} \, . \tag{23}$$

Both sides of this identity are, in the variable y, holomorphic sections of $K_C \otimes (2f) \in J_{2g-2}(C)$ with g-1 double zeroes.

Proof. For $x,a \in C$ fixed, $\frac{\Theta(y-x-f)}{\Theta(y-a-f)} \frac{E(a,y)}{E(x,y)}$ is, by (20), a meromorphic function of y with no zeroes or poles by Cor. 1.4; so setting $y = b \in C$,

$$\frac{\Theta(y-x-f)}{E(x,y)} = \frac{\Theta(y-a-f)}{E(a,y)} \cdot \frac{\Theta(b-x-f)}{\Theta(b-a-f)} \frac{E(a,b)}{E(x,b)} \qquad \forall\ x,y,a,b \in C. \tag{24}$$

Thus $\frac{\Theta(y-x-f)}{E(x,y)} = \phi(x)\psi(y)$ where ϕ and ψ are holomorphic sections in x and y; letting $y \to x$ we find $\phi(x)\psi(x) = -H_f(x)$ so that $\frac{\Theta(y-x-f)}{E(x,y)} = -H_f(x)\frac{\psi(y)}{\psi(x)}$. Now compute the Taylor expansion of both sides of this equation near $y = x$, taking y as local coordinate, and use the fact that $\left.\frac{d^2}{dy^2}E(x,y)\right|_{y=x} = 0$ from the definition (19) to conclude that

$$\psi(x) = (H_f(x))^{\frac{1}{2}} \exp\left(-\tfrac{1}{2}\int^x \frac{Q_f}{H_f}\right), \text{ which gives (23).}$$

Corollary 2.3. For $x,y \in C$ and $f \in (\Theta)$ non-singular,

$$\frac{\Theta(y-x-f)\Theta(y-x+f)}{H_f(x)H_f(y)} = E(x,y)E(y,x) = -E(x,y)^2. \tag{25}$$

When f is singular, numerator and denominator of (25) will vanish identically by Riemann's Theorem.

Corollary 2.4. Let ζ and ξ be positive divisors of degree $g-1$ such that $f = \zeta-\Delta$ and $-f = \xi-\Delta$ are non-singular points on (Θ). Then

$$\ln - \frac{\theta(y-x+f)}{\theta(y-x-f)} = \int_x^y \frac{Q_f}{H_f} = \int_x^y \omega_{\xi-\zeta} + \sum_1^g m_k \int_x^y v_k \qquad \forall\, x,y \in C$$

where $\omega_{\xi-\zeta}$ is the normalized differential of the third kind with poles of residue $+1,-1$ at ξ,ζ and $\begin{Bmatrix} m_1,\ldots,m_g \\ \ldots\ldots\ldots \end{Bmatrix}_\tau$ is the lattice point in $\mathbb{C}^g$ given by $-2f + \int_\xi^\zeta v$.

If ζ and ξ are distinct divisors, $\omega_{\xi-\zeta}$ is a meromorphic differential on C; equivalently, $Q_f(x)/H_f(x)$ will be a holomorphic differential on C if and only if f is an odd half-period. When $f \in (\Theta)$ is a singular point, $\theta(y-x-f) \equiv 0$ on $C \times C$ by Riemann's Theorem, and a Taylor expansion near $y = x$ shows that $Q_f(x) \equiv 0$ on C (Corollary 2.18).

Corollary 2.5. Suppose $x,y \in C$ have local coordinates x,y in a neighborhood of a point $p \in C$, and let $T_f(p)$ be the cubic differential $\sum_{i,j,k=1}^{g} \frac{\partial^3\theta}{\partial z_i \partial z_j \partial z_k}(f) v_i(p) v_j(p) v_k(p)$ for non-singular $f \in (\Theta)$. Then near $y = x$,

$$\frac{E(x,y)\sqrt{dx\,dy}}{y-x} = 1 - \frac{(x-y)^2}{12} S(p) + \text{higher order terms} \tag{26}$$

where

$$S(p) = \left\{\int^p H_f\,,\,p\right\} + \frac{3}{2}\left(\frac{Q_f}{H_f}\right)^2(p) - 2\,\frac{T_f}{H_f}(p) \tag{27}$$

is a *holomorphic* projective connection on C independent of the non-singular point $f \in (\Theta)$; here $\{\ ,\ \}$ is the Schwarzian differential operator - see [13, p. 164]. At a zero of $H_f(p)$, it is seen that

$$Q_f(p) = \pm H_f'(p) \quad \text{and} \quad T_f(p) = -H_f''(p) \pm \frac{3}{2} Q_f'(p)$$

with the sign ($\pm$) chosen according as $\theta(x-p\mp f) \equiv 0$ for all $x \in C$ by (25); here the derivatives ' and " have meaning in any local coordinate because of the condition on the point p.

For example, the construction on p. 13 of the prime-form for the hyperelliptic curve $s^2 = \prod_1^{2g+2} (z - z(Q_i))$ gives

$$S(p) = \{z,p\} + \frac{3}{8}\left(d\ln\prod_1^{g+1}\frac{z(p)-z(Q_{i_k})}{z(p)-z(Q_{j_k})}\right)^2 - 6\sum_{i,j=1}^{g}\frac{\partial^2 \ln\theta[\beta]}{\partial z_i\partial z_j}(0)\,v_i(p)v_j(p)$$

for *any* non-singular even half-period β corresponding to a partition $\{Q_{i_1},\dots,Q_{i_{g+1}}\} \cup \{Q_{j_1},\dots,Q_{j_{g+1}}\}$ of $\{1,2,\dots,2g+2\}$.

Differentiating (23) or (24), we obtain the fundamental normalized differential of the second kind on C:

Corollary 2.6. For $x,y \in C$,

$$\omega(x,y) = \frac{d^2}{dxdy}\ln E(x,y)dxdy = \frac{d^2}{dxdy}\ln\theta(y-x-f)dxdy \tag{28}$$

is a well-defined bilinear meromorphic differential on $C \times C$, independent of the non-singular point $f \in (\theta)$. Equivalently, for $e \in \mathbb{C}^g$ with $\theta(e) \neq 0$,

$$\omega(p,q) = -\sum_{i,j=1}^{g} v_i(p)v_j(q)\frac{\partial^2\ln\theta}{\partial z_i\partial z_j}(e) \tag{29}$$

for all $(p,q) \in C \times C$ satisfying either $\theta(p-q-e) = 0$ or $\theta(p-q+e) = 0$. (See Cor. 2.12.) In addition, $\omega(x,y)$ has the properties:

i) $\omega(x,y)$ is holomorphic everywhere except for a double pole along $y = x$ where, if x and y have local coordinates x,y in a neighborhood of $p \in C$,

$$\omega(x,y) = \left(\frac{1}{(x-y)^2} + \frac{1}{6}S(p) + \text{higher order terms}\right)dxdy$$

with S(p) the projective connection (27).

ii) For any fixed $x \in C$,

$$(30) \qquad \int_{A_j} \omega(x,y) = 0 \quad \text{and} \quad \int_{B_j} \omega(x,y) = v_j(x) \qquad j = 1,\dots,g.$$

If $\mathcal{A} = \sum_1^n a_i$ and $\mathcal{B} = \sum_1^n b_i$ are two positive divisors on C,

$$(31) \qquad \omega_{\mathcal{B}-\mathcal{A}}(x) = \int_{\mathcal{A}}^{\mathcal{B}} \omega(x,y) = d \ln \prod_1^n \frac{\theta(x-b_i-f_i)}{\theta(x-a_i-f_i)}$$

for generic non-singular $f_i \in (\theta)$; in particular

$$\int_a^b \omega_{d-c} = \int_a^b \int_c^d \omega(x,y) = \int_c^d \omega_{b-a} = \ln \frac{E(b,d)\;E(a,c)}{E(a,d)\;E(b,c)}$$

for all $a,b,c,d \in C$.

iii) The indefinite integral of $\omega(x,y)$ is an analogue of the Weierstrass ζ-function: if $Z_p(x) = \frac{d}{dx} \ln E(x,p)$ for $x,p \in C$, then $Z_p(x)$ is a meromorphic affine connection for the bundle $\delta_p^*(\theta) \otimes L_0^{-1}$ on C, where $\delta_p(x) = x-p \in J(C)$ for $x \in C$; that is, $Z_p(x)$ is a cochain in the sheaf of germs of meromorphic differentials on C with coboundary the element of $H^1(C,\Omega_C^1)$ defined by the logarithmic derivative of the transition functions for the bundle. $Z_p(x)$ has a pole of residue $+1 = \deg[\,\delta_p^*(\theta) \otimes L_0^{-1}]$ at $x = p$ and $Z_b(x) - Z_a(x) = \omega_{b-a}(x) = \int_a^b \omega(x,y)$ for distinct $a,b \in C$. For any function f on C holomorphic in some neighborhood U of p, $f(p) = \frac{1}{2\pi i} \int_{\partial U} f(x) Z_p(x)$ so that $Z_p(x)$ is a local "Cauchy-kernel".

iv) Under a change of homology basis $\begin{pmatrix} \tilde{A} \\ \tilde{B} \end{pmatrix} = \begin{pmatrix} d & c \\ b & a \end{pmatrix} \begin{pmatrix} A \\ B \end{pmatrix}$ as on p. 17 (iv), $\omega(x,y)$ becomes

$$\tilde{\omega}(x,y) = \omega(x,y) - \tfrac{1}{2} \sum_{i \le j} [v_i(x)v_j(y) + v_i(y)v_j(x)] \frac{\partial}{\partial \tau_{ij}} \ln \det(c\tau + 2\pi i d)$$

for all $x,y \in C$ so that by (12) again,

$$\omega(x,y) + \frac{2}{4^g+2^g}\sum_{i,j=1}^{g} v_i(x)v_j(y)\frac{\partial^2}{\partial z_i\partial z_j}\ln\prod_{\beta\text{ even}}\theta[\beta](z)\Big|_{z=0}$$

is a symmetric differential on $C \times C$ independent of the choice of homology defining θ and ω. Such differentials were considered by Klein in his work on Abelian functions and invariant theory [18]. By Corollary 2.12 below, this differential is also given by

$$\frac{2}{4^g+2^g}\sum_{\beta\text{ even}}\frac{\theta[\beta]^2(y-x)}{\theta[\beta]^2(0)E^2(x,y)}.$$

The theta-functions are expressed in terms of Abelian integrals by

Lemma 2.7. Let $\mathcal{A} = \sum_1^g a_i$ be a positive divisor of degree g on C with $i(\mathcal{A}) = 0$, and set $e = \mathcal{A}-a-\Delta = \int_{ga}^{\mathcal{A}} v - k^a \in \mathbb{C}^g$ with $a \in C$ and k^a given by (13). Then $\forall\ x,p \in C$

$$\text{(32)}\qquad \frac{d}{dx}\ln\theta(x-a-e) = \int_{gp}^{\mathcal{A}}\omega(x,y) - \frac{1}{2\pi i}\sum_{k=1}^{g}\int_{A_k} v_k(q)\int_p^q \omega(x,y)$$

$$\text{(33)}\qquad \frac{\theta(x-a-e)}{\theta(e)} = \exp\left\{\int_{gp}^{\mathcal{A}}\omega_{x-a} - \frac{1}{2\pi i}\sum_{k=1}^{g}\int_{A_k} v_k(q)\int_p^q \omega_{x-a}\right\}$$

$$\text{(34)}\qquad \frac{\theta(x-a-e)}{\theta(e)} = \prod_1^g\frac{E(x,a_i)}{E(a,a_i)}\exp\left\{-\frac{1}{2\pi i}\sum_{k=1}^{g}\int_{A_k} v_k(q)\ln\frac{E(x,q)}{E(a,q)}\right\}$$

and

$$\text{(35)}\qquad \frac{\partial^2}{\partial x\partial a}\ln\theta(x-a-e) = \sum_{i,j=1}^{g}\omega(x,a_i)v(\mathcal{A})^{-1}_{ij}\,v_j(a)$$

where $v(\mathcal{A})^{-1}$ is the inverse matrix to the non-singular $g \times g$ matrix $(v_i(a_j))$.

Proof. Take $\omega = \omega(x,y)$ in Prop. 1.2 (15) to get (32); (33) comes from integrating (32) between x and a, and the third equation comes from (33) and (21). Finally, (35) is obtained either by differentiating (32) or by solving the system of equations (28) with $f = a_k-a-e \in (\theta)$:

$$\omega(x,a_k) = -\sum_{i,j=1}^{g} \frac{\partial^2 \ln \Theta}{\partial z_i \partial z_j}(x-a-e)v_i(x)v_j(a_k) \qquad k = 1,\ldots,g.$$

From (33), we get an extension of Prop. 2.2 for the divisor of zeroes of meromorphic Prym differentials with double zeroes and a double pole at $y = x$:

Proposition 2.8. For $e \in \mathbb{C}^g$ with $\Theta(e) \neq 0$, let $\mathcal{A}(x)$ and $\mathcal{B}(y)$ be positive divisors of degree g such that $e = \mathcal{A}(x)-x-\Delta$ and $-e = \mathcal{B}(y)-y-\Delta$ in $J_0(C)$ for any $x,y \in C$. Then

$$\frac{\Theta^2(y-x-e)}{\Theta^2(e)} = \exp\left\{\int_{\mathcal{B}(y)}^{\mathcal{A}(x)} \omega_{y-x} + \sum_{j=1}^{g} m_j \int_x^y v_j\right\} \tag{36}$$

where $\begin{Bmatrix} m_1,\ldots,m_g \\ \cdots\cdots \end{Bmatrix}_\tau$ is the lattice point in $\mathbb{C}^g$ given by $2e - \int_{x+\mathcal{B}(y)}^{y+\mathcal{A}(x)} v$.

Proof. By (33), $\dfrac{\Theta(y-x-e_1)}{\Theta(e_1)}\dfrac{\Theta(y-x+e_2)}{\Theta(e_2)} = \exp\int_{\mathcal{A}(y)}^{\mathcal{A}(x)} \omega_{y-x}$ if $e_1 = \int_{gx}^{\mathcal{A}(x)} v - k^x$ and $e_2 = \int_{gy}^{\mathcal{A}(y)} v - k^y$ in $\mathbb{C}^g$. On the other hand,

$$\frac{\Theta(y-x-e_1)}{\Theta(y-x+e_1)} = \exp\left\{\int_{\mathcal{B}(y)}^{\mathcal{A}(y)} \omega_{y-x} + \sum_1^g \mu_i \int_x^y v_i\right\}$$

where $\begin{Bmatrix} \mu_1,\ldots,\mu_g \\ \cdots\cdots \end{Bmatrix}_\tau$ is the lattice point in $\mathbb{C}^g$ given by $2e_1 + \int_{\mathcal{A}(y)}^{\mathcal{B}(y)} v$ since, for fixed y, both sides have the same zeroes and poles and are sections of the bundle $-2e_1$ in x. Multiplying these two equations, the lattice point $\begin{Bmatrix} m_1,\ldots,m_g \\ \cdots\cdots \end{Bmatrix}_\tau$ in (36) is given by

$$2e - 2e_1 + \begin{Bmatrix} \mu_1 \cdots\cdots \mu_g \\ \cdots\cdots\cdots \end{Bmatrix}_\tau - (e_1-e_2) = 2e + \int_{\mathcal{A}(y)}^{\mathcal{B}(y)} v - (e_1-e_2) = 2e - \int_{x+\mathcal{B}(y)}^{y+\mathcal{A}(x)} v$$

Proposition 2.9. Let $v(\mathcal{A})$ be the non-singular $g \times g$ matrix $(v_i(a_j))$ for a divisor $\mathcal{A} = \sum_1^g a_i$ with $a_i \in C$ and $i(\mathcal{A}) = 0$. Then $\forall\ p,x,y \in C$

$$(37) \qquad \Lambda_p(x,y) = \frac{1}{\det v(\mathcal{A})} \det \begin{pmatrix} \int_p^y \omega(x,t) & \int_p^y \omega(a_1,t) & \cdots & \int_p^y \omega(a_g,t) \\ v_1(x) & v_1(a_1) & \cdots & v_1(a_g) \\ v_2(x) & v_2(a_1) & \cdots & v_2(a_g) \\ \vdots & \vdots & & \vdots \\ v_g(x) & v_g(a_1) & \cdots & v_g(a_g) \end{pmatrix}$$

is a well-defined differential in x and meromorphic function of y with

$$\operatorname{div}_{C\times C} \Lambda_p(x,y) > C \times \{p\} + \mathcal{A} \times C - C \times \mathcal{A} - \{p\} \times C - \text{Diagonal}$$

and

$$-\Lambda_y(x,p) = \Lambda_p(x,y) = \Lambda_p(x,q) + \Lambda_q(x,y)$$

for all $x,y,p,q \in C$. If $e \in \mathbb{C}^g$ with $e = \mathcal{A} - p - \Delta$ in $J_0(C)$, then

$$(37)' \qquad \Lambda_p(x,y) = \frac{\theta(y-x-e)\theta(x-p-e)E(p,y)}{\theta(y-p-e)\theta(e)E(x,y)E(x,p)} \qquad \forall\ x,y \in C.$$

For any smooth curve $\gamma \subset C - p$ and continuous function $\phi(x)$ on γ, $\frac{1}{2\pi i}\int_\gamma \phi(x)\Lambda_p(x,y)$ is a meromorphic function of y on $C - \gamma$ with divisor $> p - \mathcal{A}$; and if f(x) is holomorphic on the closure of some region $U \subset C$, not containing the divisor $p + \mathcal{A}$, then $f(y) = \frac{1}{2\pi i}\int_{\partial U} f(x)\Lambda_p(x,y)\ \ \forall\ y \in U$. We can thus call $\Lambda_p(x,y)$ a "Cauchy-kernel" for C associated to the divisor $p + \mathcal{A}$ - see [16, p. 651].

Proof. Corollary 2.6 and Cauchy's integral formula give all properties of $\Lambda_p(x,y)$ directly, except for (37)$'$. However, for fixed $x \in C$, $\dfrac{\Lambda_p(x,y)\theta(y-p-e)E(y,x)}{E(y,p)\theta(y-x-e)}$ is a meromorphic function of y with at most g zeroes at $\mathcal{A} = \operatorname{div}_C \theta(y-p-e)$; since $i(\mathcal{A}) = 0$, this function must be a constant $c(x,p)$ where, by (26) and (31),

$$c(x,p) = \lim_{y\to x} \frac{\theta(y-p-e)}{\theta(y-x-e)E(y,p)} \cdot \lim_{y\to x} E(y,x)\Lambda_p(x,y) = \frac{\theta(x-p-e)}{\theta(e)E(x,p)} \cdot 1.$$

The identity (37)$'$ can also be proved from (35) and the following expression for the Cauchy kernel, which gives the differentials involved in Cors. 1.3-1.4.

Proposition 2.10. For $a,b,x \in C$ and $e \in \mathbb{C}^g$:

$$(38) \quad \frac{\theta(x-a+e)\,\theta(x-b-e)}{\theta(e)\,\theta(e+b-a)} \cdot \frac{E(a,b)}{E(x,a)\,E(x,b)} = \omega_{b-a}(x) + \sum_{i=1}^{g}\left[\frac{\partial \ln\theta}{\partial z_i}(e+b-a) - \frac{\partial \ln\theta}{\partial z_i}(e)\right] v_i(x)$$

Equivalently:

$$(38)' \quad \frac{d}{dx} \ln \frac{\theta(x-a-e)\,E(x,b)}{\theta(x-b-e)\,E(x,a)} = \frac{\theta(2x-a-b-e)\,\theta(e)\,E(a,b)}{\theta(x-a-e)\,\theta(x-b-e)\,E(x,a)\,E(x,b)}$$

Proof. For fixed a and b, there is a multiplicative differential $c_{a,b}(x)$ depending on a and b such that $\forall\, z \in \mathbb{C}^g$,

$$\theta(z)\,\theta(z+b-a)\left[\omega_{b-a}(x) - \sum_{i=1}^{g} v_i(x)\frac{\partial}{\partial z_i}\ln\frac{\theta(z)}{\theta(z+b-a)}\right] = c_{a,b}(x)\,\theta(z+b-x)\,\theta(z-a+x)$$

since both sides of this equation are holomorphic sections in z of the same bundle on $J(C)$, and the left-hand side has a divisor of zeroes containing $\{x-b+f\} \sqcup \{a-x+f\}$, $\forall\, f \in (\theta)$ by (31). To evaluate $c_{a,b}(x)$ giving (38), take $z = f \in (\theta)$ non-singular; then by (24)

$$c_{a,b}(x) = \frac{-\theta(f+b-a)\,H_f(x)}{\theta(f+b-x)\,\theta(f-a+x)} = \frac{E(a,b)}{E(x,b)\,E(x,a)}$$

Finally (38)$'$ is derived by substituting $x-b-e$ for e in (38).

From (38)$'$ we see that for generic $e \in \mathbb{C}^g$, the 4g zeroes of $\theta(2x-a-b-e)$ on C are the ramification points of the (g+1)-sheeted cover of $\mathbb{P}_1(\mathbb{C})$ defined by the meromorphic function $\theta(x-a-e)E(x,b)/\theta(x-b-e)E(x,a)$.

Corollary 2.11. For generic $a,b \in C$, the 4^g differentials with 2g double zeroes and simple poles of residue -1,+1 at a,b are given by

$$\frac{\theta[e]^2(\tfrac{1}{2}\int_{a+b}^{2x} v)}{\theta[e]^2(\tfrac{1}{2}\int_a^b v)} \frac{E(a,b)}{E(x,a)E(x,b)} = \omega_{b-a}(x) + 2\sum_{i=1}^{g} \frac{\partial \ln \theta[e]}{\partial z_i}(\tfrac{1}{2}\int_a^b v)v_i(x)$$

for any half-period e (see §5, pp. 100-103). When a and $b \neq a$ are such that $\theta[e](\tfrac{1}{2}\int_a^b v) = 0$, then $\theta[e](\tfrac{1}{2}\int_{a+b}^{2x} v) \equiv 0$ on C and differentiating the above identity at b yields

$$\frac{\left(\sum_{i=1}^{g} \frac{\partial\theta[e]}{\partial z_i}(\tfrac{1}{2}\int_{a+b}^{2x} v)v_i(b)\right)^2}{\sum_{i,j=1}^{g} \frac{\partial^2\theta[e]}{\partial z_i \partial z_j}(\tfrac{1}{2}\int_a^b v)v_i(b)v_j(b)} \frac{E(a,b)}{E(x,a)E(b,x)} = \sum_{i=1}^{g} \frac{\partial\theta[e]}{\partial z_i}(\tfrac{1}{2}\int_a^b v)v_i(x).$$

This is a holomorphic differential on C vanishing at a,b and with g-2 double zeroes which are common zeroes of $\theta[e](\int_p^x v - \tfrac{1}{2}\int_a^b v)$ and $\theta[e](\int_p^x v + \tfrac{1}{2}\int_a^b v)$ for any fixed $p \in C$; thus $\forall\ x,y \in C$

$$\theta[e](\int_x^y v + \tfrac{1}{2}\int_a^b v)E(x,a)E(y,b) = -\theta[e](\int_y^x v + \tfrac{1}{2}\int_a^b v)E(y,a)E(x,b)$$

and by Cor. 2.4:

$$\sum_{i,j=1}^{g} \frac{\partial^2\theta[e]}{\partial z_i \partial z_j}(\tfrac{1}{2}\int_a^b v)v_i(x)v_j(x) = \omega_{a-b}(x) \sum_{i=1}^{g} \frac{\partial\theta[e]}{\partial z_i}(\tfrac{1}{2}\int_a^b v)v_i(x).$$

Corollary 2.12. For all $x,y \in C$ and $e \in \mathbb{C}^g$,

$$(39) \qquad \frac{\theta(y-x-e)\theta(y-x+e)}{\theta^2(e)E(x,y)^2} = \omega(x,y) + \sum_1^g \frac{\partial^2 \ln\theta}{\partial z_i \partial z_j}(e)v_i(x)v_j(y).$$

For e an even half-period, the left hand side is a differential with 2g double zeroes and a double pole at $y = x$; it is the square of the half-order differential $\frac{\theta[e](y-x)}{\theta[e](0)E(x,y)}$, a meromorphic section of L_e called the Szego kernel of C with characteristics [e]. When C is the double of a finite bordered Riemann surface R, (39) then becomes the basic relation connecting the Szego and Bergman reproducing kernels for R - see Prop. 6.14 and pp. 125-6.

Corollary 2.13. The holomorphic quartic differential

$$\frac{1}{6}\sum_{1}^{g}\frac{\partial^4 \ln\theta}{\partial z_i\partial z_j\partial z_k\partial z_\ell}(e)\,v_iv_jv_kv_\ell+\left(\sum_{1}^{g}\frac{\partial^2\ln\theta}{\partial z_i\partial z_j}(e)\,v_iv_j\right)^2+\frac{1}{3}\sum_{1}^{g}\frac{\partial^2\ln\theta}{\partial z_i\partial z_j}(e)\left(Sv_iv_j-v_iv_j''+\frac{3}{2}v_i'v_j'\right)$$

is independent of the point $e \in \mathbb{C}^g$. Here S is the projective connection (27) and v_j', v_j'' are derivatives in some local coordinate; the connection S transforms in such a way [13, p. 170] that the third term above is actually a quartic differential on C.

Proof. By computing the second order terms in the Laurent development of (39) near $y = x$ the quartic differential in question is given by $6a_3^2(x) + 6\frac{da_4}{dx}(x) - 20a_5(x)$ where $a_n(x)$ is the nth Taylor coefficient of E(x,y) at $y = x$. The differential can thus be expressed in terms of any fixed non-singular $f \in (\Theta)$; for the elliptic case, see p. 35.

From (38) and (39) we can give a special "addition-theorem" for the even Θ-functions on C:

Proposition 2.14. For $a,b,c,d \in C$

$$\frac{1}{E^2(a,b)E^2(c,d)}\left[\theta^2(d-c)\theta^2(b-a)-\theta^2(0)\theta(d-c+b-a)\theta(d-c-b+a)\right]$$

$$(40)\qquad = -\tfrac{1}{2}\theta^4(0)\sum_{i,j,k,\ell=1}^{g}\frac{\partial^4\ln\theta}{\partial z_i\partial z_j\partial z_k\partial z_\ell}(0)v_i(a)v_j(b)v_k(c)v_\ell(d)$$

$$= V(a,b,c,d) + V(a,c,b,d) + V(a,d,b,c)$$

where $V(a,b,c,d) = \frac{\theta(a-c)\theta(a-d)\theta(b-c)\theta(b-d)}{E(a,c)E(a,d)E(b,c)E(b,d)}$. In these identities,

Θ may be replaced by $\Theta[e]$ for any even half-period $e \in J_0$.

Proof. By Prop. 2.10 and the fact that $\omega_{d-c} + \omega_{c-b} = \omega_{d-b}$, the third term in (40) is:

$$\frac{\theta(0)}{E(b,c)\,E(b,d)\,E(c,d)} \sum_{i=1}^{g} v_i(a) \frac{\partial}{\partial z_i} \theta(z+d-c)\,\theta(z+c-b)\,\theta(z+b-d)\Big|_{z=0}$$

$$= \frac{\theta^2(0)\,\theta^2(d-c)}{E(c,d)\,E(d,c)} \sum_{i,j=1}^{g} \left[\frac{\partial^2 \ln\theta}{\partial z_i \partial z_j}(d-c) - \frac{\partial^2 \ln\theta}{\partial z_i \partial z_j}(0) \right] v_i(a)\, v_j(b)$$

which gives the first term of (40) by setting $e = d-c$ and $x = b$ in (39). On the other hand,

$$\frac{\partial^2 \ln\theta}{\partial z_i \partial z_j}(d-c) = \frac{1}{2}\, \frac{E^2(c,d)}{\theta^2(d-c)}\, \frac{\partial^2}{\partial z_i \partial z_j} \left[\frac{\theta(z+d-c)\,\theta(z+c-d)}{E^2(c,d)} \right]_{z=0}$$

so using Cor. 2.12 again, we get the second term of (40).

The symmetric holomorphic differential in (40) figures into the Schiffer variation of the Szego kernel - see [15]. Letting $c = a$ and $d = b$, we get

Corollary 2.15. For any even half period $e \in J_0(C)$ and $\forall\, x,y \in C$,

$$\frac{\theta[e]\left(2\int_x^y v\right)}{\theta[e](0)\, E^4(x,y)} - \frac{\theta[e]^4\left(\int_x^y v\right)}{\theta[e]^4(0)\, E^4(x,y)} = \frac{1}{2} \sum_{i,j,k,l=1}^{g} \frac{\partial^4 \ln \theta[e]}{\partial z_i \partial z_j \partial z_k \partial z_l}(0)\, v_i(x)\, v_j(x)\, v_k(y)\, v_l(y)$$

Comparing the second-order terms in the Laurent development of this identity near $y = x$ gives an expression for the projective connection of Cor. 2.5 in terms of any even theta function - for the elliptic case, see p. 36. Though there is no analogue of this corollary for odd half-periods, Cor. 2.12 does imply

$$\frac{\theta(2y-2x-f)}{E^2(x,y)} = \frac{\theta^2(y-x-f)}{H_f(p)} \sum_{i,j,k=1}^{g} \frac{\partial^3 \ln\theta}{\partial z_i \partial z_j \partial z_k}(y-x-f)\, v_i(x)\, v_j(y)\, v_k(p)$$

for all non-singular $f \in (\theta)$ and $p,x,y \in C$. Likewise (39) gives a "duplication formula" for generic $e \in \mathbb{C}^g$:

$$(41)\quad \theta(2y-2x-e)\theta(e) = -E^2(x,y)\theta(y-x-e)^2 \frac{\partial^2}{\partial x\partial y} \ln \frac{\theta(y-x-e)}{E(x,y)} \quad \forall\, x,y \in C.$$

These identities are all special cases of a general addition theorem (42)-(43) which we now take up.

Proposition 2.16.* Let D be a divisor of degree $g+n-1 \geq g-1$ on C with $\dim H^0(D) = N \geq n$, and suppose $\psi_1,\ldots,\psi_N$ form a basis for the holomorphic sections of the line bundle corresponding to D as on p. 5. Then for any (generic) positive divisor $\mathcal{B} = \sum_1^{N-n} b_i$ with $i(D+\mathcal{B}) = 0$ and for $(x_1,\ldots,x_N) \in C\times C\times\ldots\times C = C^N$,

$$(42)\quad \operatorname{div}_{C^N}\theta\Big(\sum_1^N x_i + \Delta - D - \mathcal{B}\Big) = \operatorname{div}_{C^N} \frac{\det(\psi_i(x_j))}{\prod_{i<j} E(x_i,x_j)} \prod_{i=1}^{N}\prod_{j=1}^{N-n} E(x_i,b_j).$$

Proof. The divisors $\mathcal{B} = \sum_1^{N-n} b_i$ with $i(D+\mathcal{B}) > 0$ are a subvariety of codimension one in C^{N-n} given by the zeroes of $\det(V_i(b_j))$ for any basis $V_1,\ldots,V_{N-n}$ of the Abelian differentials with divisor $\geq D$; so the divisors $\mathcal{B}$ with $i(D+\mathcal{B}) = 0$ are generic in C^{N-n} and for such $\mathcal{B}$, $\theta(\sum_1^N x_i + \Delta - D - \mathcal{B}) \not\equiv 0$ on C^N since otherwise the condition $\dim H^0(D+\mathcal{B} - \sum_1^N x_i) > 0$ for all $x_1,\ldots,x_N \in C$ would imply $\dim H^0(D+\mathcal{B}) > N$. Now since $\dim H^0(D) = N$, there is, for any positive divisor Σ of degree $N-1$, a positive divisor $\mathcal{A}$ of degree $g+n-N$ such that $\mathcal{A} + \Sigma = D$ in J_{g+n-1}. Consequently, for generic $x_2,\ldots,x_N \in C$, $\operatorname{div}_C\theta(\sum_1^N x_i + \Delta - D - \mathcal{B}) = \mathcal{B} + \mathcal{A}$ as a function of x_1, where $\mathcal{A}$ is the unique positive divisor of degree $g+n-N$ for which $\mathcal{A} + \sum_2^N x_i = D$. So the section in x_1 given by

---

* This theorem has appeared in various forms classically - see [7], [18], [20] or [30].

$$\theta\Big(\sum_1^N x_i + \Delta - D - \mathcal{B}\Big) \prod_{i<j} E(x_i,x_j) \Big/ \prod_{i=1}^{N} \prod_{j=1}^{N-n} E(x_i,b_j)$$

is a holomorphic section of a line bundle holomorphically equivalent to $D \in J_{g+n-1}$ with zeroes at $\mathcal{A} + \sum_2^N x_i$, and must therefore have the same divisor of zeroes as $\det(\psi_i(x_j))$, a holomorphic section in x_1 of $D \in J_{g+n-1}$ with zeroes at $\sum_2^N x_i$. Since both the section and determinant are symmetric in $x_1,\ldots,x_N$, this argument proves (42).

For example (Riemann), if $\xi \in J_0$ and $\psi_1,\ldots,\psi_{2g-2}$ are a basis for the holomorphic sections of $K_C \otimes L_0 \otimes \xi \in J_{3g-3}(C)$, then

$$\operatorname{div}_{C^{2g-2}} \theta\Big(\sum_1^{2g-2} x_i - K_C - \xi\Big) = \operatorname{div}_{C^{2g-2}} \frac{\det(\psi_i(x_j))}{\prod_{i<j} E(x_i,x_j)} .$$

Likewise if $0 \neq \xi \in J_0$ and $\phi_1,\ldots,\phi_{g-1}$ are a basis for the holomorphic sections of $K_C \otimes \xi \in J_{2g-2}$,

$$\operatorname{div}_{C^{g-1}} \theta\Big(\sum_1^{g-1} x_i - \Delta - \xi\Big) = \operatorname{div}_{C^{g-1}} \frac{\det(\phi_i(x_j))}{\prod_{i<j} E(x_i,x_j)} ,$$

while if $\phi_1,\ldots,\phi_g$ are a basis for the meromorphic sections of $K_C \otimes \xi \in J_{2g-2}$ with at most a simple pole at some fixed point $p \in C$,

$$\operatorname{div}_{C^g} \theta\Big(\sum_1^g x_i - p - \Delta - \xi\Big) = \operatorname{div}_{C^g} \frac{\det(\phi_i(x_j))}{\prod_{i<j} E(x_i,x_j)} \prod_1^g E(x_i,p)$$

When $\xi = 0$, these last equations change to:

<u>Corollary 2.17</u>. For all $p,x_1,\ldots,x_g \in C$,

$$\frac{\theta\big(\sum_1^g x_i - p - \Delta\big)}{\prod_1^g E(x_i,p)} = c\,\frac{\det(v_i(x_j))}{\prod_{i<j} E(x_i,x_j)}\,\frac{\sigma(p)}{\prod_1^g \sigma(x_i)}$$

where c is a constant independent of $p, x_1, \ldots, x_g$ and

$$\sigma(p) = \exp\left\{-\frac{1}{2\pi i}\sum_{j=1}^{g}\int_{A_j} v_j(y)\ln E(y,p)\right\}$$

is a holomorphic section of a trivial bundle on C. In particular, if $f = \sum_1^{g-1} a_i - \Delta \in (\Theta)$ with Δ given by (13)-(14), then

$$c \det\begin{pmatrix} v_1(x) & v_1(a_1) & \cdots & v_1(a_{g-1}) \\ \vdots & \vdots & & \vdots \\ v_g(x) & v_g(a_1) & \cdots & v_g(a_{g-1}) \end{pmatrix} = (-1)^g \prod_{\ell} \sigma(a_\ell) \prod_{i<j} E(a_i,a_j) \sum_{k=1}^{g} \frac{\partial\theta}{\partial z_k}(f)\, v_k(x)$$

<u>Proof</u>. Prop. 2.16 implies that $\forall\ p, x_1, \ldots, x_g \in C$,

$$\frac{\theta(\sum_1^g x_i - p - \Delta)}{\prod_1^g E(x_i,p)} = c(\mathcal{A},\mathcal{B}) \frac{\det(v_i(x_j))}{\prod_{i<j} E(x_i,x_j)} \frac{\theta(\sum_1^g a_i - p - \Delta)}{\prod_1^g E(a_i,p)} \prod_{i=1}^{g} \frac{\prod_{j=1}^{g} E(b_j,x_i)}{\theta(\sum_1^g b_j - x_i - \Delta)}$$

where $c(\mathcal{A},\mathcal{B})$ depends only on chosen fixed divisors $\mathcal{A} = \sum_1^g a_i$ and $\mathcal{B} = \sum_1^g b_i$. On the other hand, from (34) of Lemma 2.7,

$$\theta(\sum_1^g x_i - p - \Delta) = s(x_1,\ldots,x_g)\sigma(p) \prod_{i=1}^{g} E(x_i,p) \qquad \forall\ p \in C$$

where $s(x_1,\ldots,x_g)$ is a symmetric holomorphic section of a line bundle on $C \times \ldots \times C$ of degree $g-1$ in each variable. Combining these two identities gives the corollary with the universal constant $c = c(\mathcal{A},\mathcal{B})s(a_1,\ldots,a_g)[s(b_1,\ldots,b_g)]^{-g}$.

By specializing $x_1 = \ldots = x_g = x \in C$, this corollary tells us that $\operatorname{div}_C \theta(gx-p-\Delta) = gp + W$ where W, independent of $p \in C$ and the choice of homology defining θ, is of degree $g^3 - g$ and is the divisor of Weierstrass points on C [13, p. 123] counted with their "weights", given by the order of the zero of $\theta(gx-p-\Delta)$ for $x \in W$. Likewise if $\xi \neq 0$ in J_0, $\operatorname{div}_C \theta((g-1)x - \Delta - \xi)$ is of degree $g(g-1)^2$ and gives the Weierstrass points for $K_C \otimes \xi$ consisting of the zeroes of the Wronskian for any basis of $H^0(K_C \otimes \xi)$.

Corollary 2.18. Let $f \in (\Theta)$ be a singular point of order m. Then for any two positive divisors $\mathcal{A} = \sum_1^m a_i$ and $\mathcal{B} = \sum_1^m b_i$ of degree m on C,

$$\frac{\theta(\sum_1^m x_i - \mathcal{A} + f)\theta(\sum_1^m x_i - \mathcal{B} - f)}{\theta(\mathcal{A} - \mathcal{B} - f)H_f(x_1,\dots,x_m)} = (-1)^{\frac{1}{2}m(m+1)} \frac{\prod_{i,j=1}^m E(x_i,a_j)E(x_i,b_j)}{\prod_{i,j=1}^m E(a_i,b_j)\prod_{i<j} E^2(x_i,x_j)}$$

where, for any $x_1,\dots,x_m \in C$,

$$H_f(x_1,\dots,x_m) = \sum_{i_1,\dots,i_m=1}^{g} \frac{\partial^m \theta}{\partial z_{i_1}\cdots\partial z_{i_m}}(f)v_{i_1}(x_1)\dots v_{i_m}(x_m).$$

Thus for all $x_i, y_j \in C$,

$$\frac{\theta(\sum_1^m (x_i-y_i)+f)\theta(\sum_1^m (x_i-y_i) - f)}{H_f(x_1,\dots,x_m)H_f(y_1,\dots,y_m)} = (-1)^m \frac{\prod_{i,j=1}^m E^2(x_i,y_j)}{\prod_{i<j} E^2(x_i,x_j)E^2(y_i,y_j)}$$

and the differential $H_f(x_1,\dots,x_m)$ vanishes identically to the second order when $x_i = x_j$ for some $i \neq j$.

Proof. Setting $D = \Delta + f$, $N = m$ and $n = 0$ in Prop. 2.16, there is a constant c depending on $\mathcal{A}$ and $x_1,\dots,x_m$ such that for all $y_1,\dots,y_m \in C$,

$$\theta(\sum_1^m y_i - \sum_1^m x_i + f) = c\,\theta(\sum_1^m y_i - \mathcal{A} + f)\prod_{i,j=1}^m \frac{E(x_i,y_j)}{E(a_i,y_j)} \quad *$$

Differentiating with respect to $y_1,\dots,y_m$ and setting $y_i = x_i$,

$$\sum_1^g \frac{\partial^m \theta}{\partial z_{i_1}\cdots\partial z_{i_m}}(f)v_{i_1}(x_1)\dots v_{i_m}(x_m) = c\,\theta(\sum_1^m x_i - \mathcal{A} + f)\prod_{j=1}^m \frac{\prod_{i\neq j} E(x_i,x_j)}{\prod_{i=1}^m E(a_i,x_j)}$$

---

* This is implicitly used in classical proofs of Riemann's Theorem - see [19, p. 434].

which gives the corollary by evaluating c using the divisor $\sum_1^m y_i = \mathcal{B}$

From Prop. 2.16 we can prove an "addition-theorem" for Abelian functions:

Corollary 2.19. If $e \in \mathbb{C}^g$ with $\Theta(e) \neq 0$,

$$(43)\quad \Theta\Big(\sum_1^n x_i - \sum_1^n y_i - e\Big)\Theta(e)^{n-1} \frac{\prod_{i<j} E(x_i,x_j)E(y_j,y_i)}{\prod_{i,j} E(x_i,y_j)} = \det\left(\frac{\Theta(x_i-y_j-e)}{E(x_i,y_j)}\right)$$

for all $x_1,\ldots,x_n,y_1,\ldots,y_n \in C$. If $f \in \mathbb{C}^g$ is a non-singular point of (Θ), then

$$\Theta\Big(\sum_1^n x_i - \sum_1^n y_i - f\Big)H_f(p)^{n-1} \frac{\prod_{i<j} E(x_i,x_j)E(y_j,y_i)}{\prod_{i=1}^n \Theta(x_i-y_i-f)\prod_{i\neq j} E(x_i,y_j)} =$$

$$(-1)^{n-1}\frac{d}{dt}\det\left(t + \sum_{k=1}^g \frac{\partial \ln\Theta}{\partial z_k}(x_i-y_j-f)v_k(p)\right)_{t=0} \qquad \forall\, p \in C.$$

In particular, specializing $y_1 = \ldots = y_n = a \in C$,

$$(44)\quad \Theta\Big(\sum_1^n (x_i-a)-f\Big)\frac{H_f(p)^{n-1}\prod_{i<j} E(x_i,x_j)}{\prod_{i=1}^n \Theta(x_i-a-f)E(x_i,a)^{n-1}} = \frac{(-1)^{\frac{1}{2}(n-1)(n-2)}}{\prod_{k=1}^{n-1} k!}\det(\phi_i(x_j))$$

where, for $j = 1,\ldots,n$, $\phi_1(x_j) = 1$ and

$$\phi_i(x_j) = \frac{\partial^{i-1}}{\partial y^{i-1}}\left(\sum_{k=1}^g \frac{\partial \ln\Theta}{\partial z_k}(x_j-y-f)v_k(p)\right)_{y=a}.$$

Proof. If $\Theta(e) \neq 0$, take $D = \Delta + e + \sum_1^n y_i$, $N = n$ and sections $\psi_k(x) = \frac{\Theta(x-y_k-e)}{E(x,y_k)}\prod_{i=1}^n E(x,y_i)$ in (42). For a non-singular $f \in (\Theta)$, take $e = f+p-q$ in (43) for $p,q \in C$ and let $q \to p$, making use of the factorization of Prop. 2.2.

Many relations discussed in this chapter can be considered specializations of the case $n = 2$ of (43):

(45) $\Theta(x-a-e)\Theta(y-b-e)E(x,b)E(a,y) + \Theta(x-b-e)\Theta(y-a-e)E(x,a)E(y,b)$

$$= \Theta(x+y-a-b-e)\Theta(e)E(x,y)E(a,b) \qquad \forall\, x,y,a,b \in C$$

and this in turn can be used to obtain the addition-theorem (43) for arbitrary n. In the elliptic case, (44) is the classical determinant form of the addition theorem for Weierstrass' $\wp$-function:

Example. If C has genus 1, the universal cover of $C \approx J(C)$ can be taken to be the z-plane $\mathbb{C}$, and the normalized holomorphic differential will then be given by $v(x) = dz(x)$, $x \in C$, with period matrix $(2\pi i, \tau)$, Re $\tau < 0$. There are four Θ-functions with half-integer characteristics: $\begin{bmatrix} 0 \\ 0 \end{bmatrix}$, $\begin{bmatrix} 0 \\ \frac{1}{2} \end{bmatrix}$ and $\begin{bmatrix} \frac{1}{2} \\ 0 \end{bmatrix}$ even and $\begin{bmatrix} \frac{1}{2} \\ \frac{1}{2} \end{bmatrix}$ odd; the class $\Delta = \left\{ \begin{matrix} \frac{1}{2} \\ \frac{1}{2} \end{matrix} \right\}_\tau = \frac{2\pi i + \tau}{2} \in J_0(C)$ is the only point on (Θ). The Weierstrass functions $\wp(z)$, $\zeta(z)$ and $\sigma(z)$ constructed from the lattice in $\mathbb{C}$ generated by $2\pi i$ and τ satisfy the relations:

$$\wp(z) = -\frac{d^2}{dz^2} \ln \sigma(z) = -\frac{d^2}{dz^2} \ln \Theta \begin{bmatrix} \frac{1}{2} \\ \frac{1}{2} \end{bmatrix}(z) + \eta$$

(46) $$\zeta(z) = -\int^z \wp(z) = \frac{d}{dz} \ln \sigma(z) = \frac{d}{dz} \ln \Theta \begin{bmatrix} \frac{1}{2} \\ \frac{1}{2} \end{bmatrix}(z) - \eta z$$

$$\frac{\sigma(z+2\pi i)}{\sigma(z)} = -e^{-2\pi i \eta(z+\pi i)} \quad \text{and} \quad \frac{\sigma(z+\tau)}{\sigma(z)} = -e^{-(\tau\eta+1)(z+\frac{1}{2}\tau)}$$

where

$$\eta = \frac{1}{3} \frac{\Theta''' \begin{bmatrix} \frac{1}{2} \\ \frac{1}{2} \end{bmatrix}(0)}{\Theta' \begin{bmatrix} \frac{1}{2} \\ \frac{1}{2} \end{bmatrix}(0)} = \frac{1}{2\pi i} \int_A \wp(z) dz = \frac{1}{2\pi i} \left(\zeta(z) - \zeta(z+2\pi i) \right)$$

The connection S(p) of Corollary 2.5 is the quadratic differential $-6\eta(dz)^2$, and for $x,y \in C$, the prime-form is given by

$$E(x,y)\sqrt{dz(x)dz(y)} = \frac{\theta\begin{bmatrix}\frac12\\ \frac12\end{bmatrix}(y-x)}{\theta\begin{bmatrix}\frac12\\ \frac12\end{bmatrix}'(0)} = \sigma\Big(\int_x^y dz\Big)\exp\frac{\eta}{2}\Big(\int_x^y dz\Big)^2 .$$

The differential of the second kind on $C \times C$ is

$$\omega(x,y) = \frac{d^2}{dx\,dy}\ln\theta\begin{bmatrix}\frac12\\ \frac12\end{bmatrix}(y-x)\,dxdy = \Big[\wp\Big(\int_x^y v\Big) - \eta\Big]v(x)v(y) ,$$

so that the "invariant" differential of Cor. 2.6 (iv) is given by

$$\frac13\sum_{\beta\text{ even}}\frac{\theta[\beta]^2\big(\int_x^y v\big)}{\theta[\beta]^2(0)\,E^2(x,y)} = \omega(x,y) + \frac13\sum_{\beta\text{ even}}\frac{\partial^2}{\partial z^2}\ln\theta[\beta](z)\Big|_{z=0} v(x)v(y)$$

$$= \omega(x,y) + \frac13\,\frac{\theta'''\begin{bmatrix}1/2\\ 1/2\end{bmatrix}(0)}{\theta'\begin{bmatrix}1/2\\ 1/2\end{bmatrix}(0)}\,v(x)v(y) = \wp\Big(\int_x^y v\Big)\,dz(x)\,dz(y)$$

by virtue of Jacobi's identity [19, p. 334] and the "heat equation" (52) below:

$$\frac{\partial^2}{\partial z^2}\ln\prod_{\beta\text{ even}}\theta[\beta](z)\Big|_{z=0} = 2\frac{\partial}{\partial T}\ln\theta'\begin{bmatrix}1/2\\ 1/2\end{bmatrix}(0) = \frac{\partial^2}{\partial z^2}\ln\theta'\begin{bmatrix}1/2\\ 1/2\end{bmatrix}(z)\Big|_{z=0}$$

For any even half-period e, Corollary 2.12 implies that $\left(\wp(z) - \eta + \frac{\theta''[e](0)}{\theta[e](0)}\right)dz$ is a differential with a double pole at $z = 0$ and double zero at $z = e + \Delta = e + \begin{Bmatrix}\frac12\\ \frac12\end{Bmatrix}_T$; this implies the well-known formula:

$$\wp(e+\Delta) = \frac13\,\frac{\theta'''\begin{bmatrix}\frac12\\ \frac12\end{bmatrix}(0)}{\theta'\begin{bmatrix}\frac12\\ \frac12\end{bmatrix}(0)} - \frac{\theta''[e](0)}{\theta[e](0)}$$

where $e + \Delta$ is a non-zero half-period for even e. The quartic differential in Corollary 2.13 is $(dz)^4$ times the constant

$$\frac16\,\frac{d^4\ln\theta}{dz^4}(z) + \left(\frac{d^2\ln\theta}{dz^2}(z)\right)^2 - 2\eta\,\frac{d^2\ln\theta}{dz^2}(z) = \frac16\left(9\eta^2 - \frac{\theta^{(v)}\begin{bmatrix}1/2\\ 1/2\end{bmatrix}(0)}{\theta'\begin{bmatrix}1/2\\ 1/2\end{bmatrix}(0)}\right)$$

for all $z \in \mathbb{C}$. The special addition theorem (40) is, for β an even half-period and $z,w \in \mathbb{C}$,

$$\theta[\beta](z+w)\,\theta[\beta](z-w)\,\theta[\beta]^2(0) - \theta[\beta]^2(z)\,\theta[\beta]^2(w)$$

(40)'

$$= \theta\begin{bmatrix}1/2\\1/2\end{bmatrix}^2(z)\,\theta\begin{bmatrix}1/2\\1/2\end{bmatrix}^2(w)\,\frac{\theta[\beta]^4(0)}{2\theta'\begin{bmatrix}1/2\\1/2\end{bmatrix}^4(0)}\,\frac{d^4}{dz^4}\ln\theta[\beta](0)$$

while Cor. 2.15 gives the connection $S(p)$ in terms of any even half-period β as $S(p) = -(4c_2 + \frac{c_6}{2c_4})dz(p)^{\otimes 2}$ where $c_n = \frac{d^n \ln \theta[\beta]}{dz^n}(0)$ and $c_4 \neq 0$ for any τ by (40)'. The duplication formula at the bottom of p. 28 becomes:

$$\sigma(2z) = \sigma^4(z)\,\frac{d^3}{dz^3}\ln\sigma(z) = -\sigma^4(z)\,\wp'(z)$$

and the addition theorem (44) is:

$$\det\begin{pmatrix} 1 & \wp(z_1) & \cdots & \wp^{(n-2)}(z_1) \\ \vdots & \vdots & & \vdots \\ 1 & \wp(z_n) & \cdots & \wp^{(n-2)}(z_n)\end{pmatrix} = (-1)^{\frac{(n-1)(n-2)}{2}} \prod_{k=1}^{n-1} k!\ \frac{\sigma(z_1+\cdots+z_n)\prod_{i<j}\sigma(z_i-z_j)}{\prod_1^n \sigma^n(z_j)}$$

for all $z_1,\ldots,z_n \in \mathbb{C}$.

III. Degenerate Riemann Surfaces

The variational method is an important tool for evaluating universal constants on the moduli space as well as for generating relations between the theta-functions and Abelian differentials on a Riemann surface. In this chapter, variational formulas are worked out for the two basic types of degenerating moduli obtained by "pinching" either a zero or non-zero homology cycle on a Riemann surface.

Pinching a Cycle Homologous to Zero. Let $\mathcal{C} \to D$ be a family of Riemann surfaces over the unit disc $D = \{t \in \mathbb{C} \mid |t| < 1\}$ constructed as follows: take two Riemann surfaces C_1 and C_2 of genus g_1, g_2 each with a point p_1, p_2 removed and let $z_1 : U_1 \tilde{\to} D$ and $z_2 : U_2 \tilde{\to} D$ be coordinates in neighborhoods U_1, U_2 of these points. For $k = 1,2$ set

$$W_k = \{(x_k, t) \mid t \in D, \quad x_k \in C_k - U_k \quad \text{or} \quad x_k \in U_k \quad \text{with} \quad |z_k(x_k)| > |t|\}$$

and let S be the non-singular surface $\{XY = t \mid (X,Y,t) \in D \times D \times D\}$. Then $\mathcal{C} = W_1 \cup S \cup W_2$ where, in the overlaps,

$$(x_1, t) \in W_1 \cap U_1 \times D \quad \text{is identified with} \quad (z_1(x_1), \frac{t}{z_1(x_1)}, t) \in S$$

and

$$(x_2, t) \in W_2 \cap U_2 \times D \quad \text{is identified with} \quad (\frac{t}{z_2(x_2)}, z_2(x_2), t) \in S.$$

Choose coordinates $x = \frac{1}{2}(X+Y)$ and $y = \frac{1}{2}(X-Y)$ on S so that the fibers C_t of $\mathcal{C}$ are Riemann surfaces of genus $g = g_1 + g_2$ for which the "pinched regions" $C_t \cap S$ are ramified double coverings $y = \sqrt{x^2 - t}$ of a neighborhood of $x = 0$ with branch points at $x = \pm\sqrt{t}$. For $t = 0$, the fiber C_0 crosses itself at the point p corresponding to $x = y = t = 0$, and the two components C_1 and C_2 of the normalization of C_0 have, on $C_0 \cap S$, the equations $y = x$ and $y = -x$ respectively; the corresponding local uniformizing variables $x = \frac{1}{2}z_1$ and $x = \frac{1}{2}z_2$ will be called the pinching coordinates for C_1 and C_2 at $p = p_1, p_2$ respectively. For each t, a canonical basis $A_1(t), B_1(t), \ldots, A_g(t), B_g(t)$ of $H_1(C_t, \mathbb{Z})$ can be chosen by extending across $(C_1 - U_1) \times D$ and $(C_2 - U_2) \times D$ the homology bases for C_1 and C_2 lying in $C_1 - U_1$ and $C_2 - U_2$ respectively.

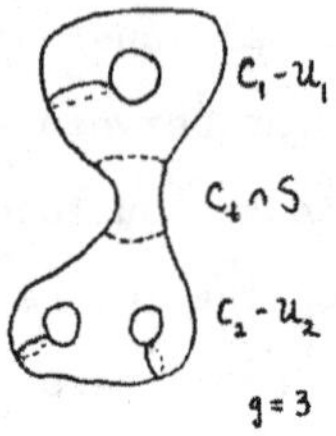

Proposition 3.1. There are g_1+g_2 linearly independent holomorphic 2-forms on $\wp$ whose residues $u_1(x,t), \ldots, u_{g_1+g_2}(x,t)$ along C_t are a normalized basis of the holomorphic differentials on C_t for t in a sufficiently small disc D_ε of radius ε about $t = 0$. For $i = 1, 2, \ldots, g_1$ and $j = g_1+1, \ldots, g_1+g_2$,

$$(47)\qquad u_i(x,t) = \begin{cases} v_i(x) + \frac{1}{4}tv_i(p_1)\omega_1(x,p_1) + O(t^2) & x \in C_1 - U_1 \\ \frac{1}{4}tv_i(p_1)\omega_2(x,p_2) + O(t^2) & x \in C_2 - U_2 \end{cases}$$

$$u_j(x,t) = \begin{cases} \frac{1}{4}tv_j(p_2)\omega_1(x,p_1) + O(t^2) & x \in C_1 - U_1 \\ v_j(x) + \frac{1}{4}tv_j(p_2)\omega_2(x,p_2) + O(t^2) & x \in C_2 - U_2 \end{cases}$$

where $v_i(x)$ for $i \leq g_1$ (resp. $v_j(x)$ for $j > g_1$) are a normalized basis for the holomorphic differentials on C_1 (resp. C_2), $\omega_1(x,y)$ and $\omega_2(x,y)$ are the normalized differentials of the second kind (28) on $C_1 \times C_1$ and $C_2 \times C_2$, and each term $O(t^2)$ is a holomorphic differential on W_1 or W_2, for which $\lim_{t \to 0} \frac{1}{t^2} O(t^2)$ is a meromorphic differential on C_1 or C_2 with at most a pole of order 4 at p. The differentials $v_i(p_1)$, $v_j(p_2)$ and $\omega_k(x,p_k)$ are all evaluated in terms of the pinching coordinates.

Proof.* If $\Omega^2_{\mathcal{C}}$ is the sheaf of holomorphic 2-forms on $\mathcal{C}$ and Ω_{C_a} is the sheaf of holomorphic differentials on C_a for $a \in D$, then on $\mathcal{C}$ there is the exact sequence of sheaves

$$0 \longrightarrow \Omega^2_{\mathcal{C}} \xrightarrow{m_a} \Omega^2_{\mathcal{C}} \xrightarrow{r_a} \Omega_{C_a} \longrightarrow 0$$

where m_a is multiplication by $(t-a)$ and r_a is the residue mapping. This in turn gives rise to the exact sequence on D:

$$0 \to \pi_*\Omega^2_{\mathcal{C}} \xrightarrow{m_a} \pi_*\Omega^2_{\mathcal{C}} \xrightarrow{r_a} \pi_*\Omega_{C_a} \longrightarrow \pi_* H^1(\Omega^2_{\mathcal{C}}) \xrightarrow{m_a} \pi_* H^1(\Omega^2_{\mathcal{C}})$$

$$\xrightarrow{r_a} \pi_* H^1(\Omega_{C_a}) \longrightarrow \pi_* H^2(\Omega^2_{\mathcal{C}}) \xrightarrow{m_a} \pi_* H^2(\Omega^2_{\mathcal{C}}) \to 0$$

Since $\Omega^2_{\mathcal{C}}$ is locally free, the direct image sheaves $\mathcal{E}^k = \pi_* H^k(\Omega^2_{\mathcal{C}})$, $k \geq 0$, are all coherent on D by a theorem of Grauert [12, p. 59]; this implies that in any sufficiently small neighborhood $U \subseteq D$, the vector space $H^0(\mathcal{E}^k|U)$ is a finitely generated $H^0(\mathcal{O}|U)$-module [14, p. 27] so that $\mathcal{E}^k$ will be a locally free sheaf on U if and only if $m_a: H^0(\mathcal{E}^k|U) \to H^0(\mathcal{E}^k|U)$ is injective $\forall\, a \in U$. Now in the above exact sequence, $\pi_* H^2(\Omega^2_{\mathcal{C}})$ is the zero sheaf since m_a cannot be surjective $\forall$ a; thus $\mathcal{E}^1/m_a\mathcal{E}^1 \cong \pi_* H^1(\Omega_{C_a})$, a skyscraper sheaf on D with stalk 0 at $t \neq a \in D$ and $\mathbb{C} = H^1(C_a, \Omega_{C_a})$ at $t = a$. Since a is arbitrary, $\mathcal{E}^1$ is a locally free sheaf of rank $1 = \dim H^1(\Omega_{C_a})$ on D and $r_a: \pi_*\Omega^2_{\mathcal{C}} \to \pi_*\Omega_{C_a}$ is surjective $\forall\, a \in D$, with $\pi_*\Omega^2_{\mathcal{C}}$ a locally free sheaf of rank $g = \dim H^0(C_a, \Omega_{C_a})$. There exist, then, holomorphic two forms $U_1(x,t), \ldots, U_{g_1+g_2}(x,t)$ for t near 0 whose residues along $t = 0$ are a given normalized basis $v_1, \ldots, v_{g_1}, v_{g_1+1}, \ldots, v_{g_1+g_2}$ of the differentials on C_0. Since $\left(\frac{1}{2\pi i} \int_{A_j} \mathrm{Res}_{C_t} U_i(x,t) \right)_{1 \leq i,j \leq g}$ is a

---

* Suggested by D. Mumford.

holomorphic matrix which is the identity matrix at $t = 0$, it is invertible in a neighborhood D_ε of $t = 0$; changing the basis $\{U_n,\ n = 1,\dots,g\}$ by this matrix, and taking the residues along C_t we then get the normalized basis $u_1(x,t),\dots,u_g(x,t)$ for the differentials on C_t. Now each differential $u_i(x,t)$ has an expansion in a neighborhood of $x = 0$ in $C_t \cap S$ given in terms of the pinching coordinate x by :

$$u_i(x,t) = \sum_{\mu\geq 0} a_\mu(t)x^\mu dx + \sum_{\nu\geq 0} b_\nu(t)\,\frac{x^\nu}{\sqrt{x^2-t}}\,dx$$

where a_μ and b_ν are holomorphic functions near $t = 0$; for $i \leq g_1$,

$$u_i(x,0) = b_0(0)x^{-1}dx + \sum_{\mu\geq 0}\Big(a_\mu(0) + b_{\mu+1}(0)\Big)x^\mu dx = v_i(x), \qquad x \in C_1 \cap S$$

and

$$u_i(x,0) = -b_0(0)x^{-1}dx + \sum_{\mu\geq 0}\Big(a_\mu(0) - b_{\mu+1}(0)\Big)x^\mu dx = 0, \qquad x \in C_2 \cap S$$

so that $b_0(0) = 0$ and $b_1(0) = a_0(0) = \frac{1}{2}v_i(p_1)$. Hence, for $x \in C_1 \cap S$,

$$\lim_{t\to 0} \frac{1}{t}\Big(u_i(x,t) - u_i(x,0)\Big) = \sum_{\mu\geq 0} a_\mu'(0)x^\mu dx + \sum_{\nu\geq 0}\Big(\frac{1}{2}b_{\nu+1}(0) + x b_\nu'(0)\Big)x^{\nu-2}dx$$

is a differential of the second kind on C_1 with at most a double pole at $x = p_1$ which has zero A-periods since $u_i(x,t)$ and $u_i(x,0)$ are normalized differentials and which, in terms of the pinching coordinates, has leading coefficient $\frac{1}{2}b_1(0) = \frac{1}{4}v_i(p_1)$ in its Laurent development at p_1; from the properties of $\omega(x,y)$ given in Cor. 2.6, we see that this differential must actually be $\frac{1}{4}v_i(p_1)\omega_1(x,p_1)$. The other expansions are similarly proved, and the assertion concerning the terms $O(t^2)$ comes from the second-order terms of the Taylor-developments of $u_i(x,t)$ at $t = 0$.

<u>Corollary 3.2</u>. The Riemann matrix $\tau(t)$ for C_t has an expansion

$$\tau(t) = \begin{pmatrix} \tau_1 & 0 \\ 0 & \tau_2 \end{pmatrix} + \tfrac{1}{4}t\, R\cdot R + O(t^2) \tag{48}$$

where τ_1 and τ_2 are the Riemann matrices for C_1 and C_2,

$$R = \left(v_1(p_1),\dots,v_{g_1}(p_1),v_{g_1+1}(p_2),\dots,v_{g_1+g_2}(p_2)\right) \in \mathbb{C}^g$$

and $O(t^2)$ is a matrix of constants with $\lim\limits_{t\to 0} \frac{1}{t^2} O(t^2)$ a finite matrix. The canonical differential $\omega_t(x,y)$ for C_t has an expansion

$$\omega_t(x,y) = \begin{cases} \omega_1(x,y) + \tfrac{1}{4}t\omega_1(x,p)\omega_1(y,p) + o(t) & x,y \in C_1-U_1 \\ \tfrac{1}{4}t\omega_1(x,p)\omega_2(y,p) + o(t) & x \in C_1-U_1,\ y \in C_2-U_2 \\ \omega_2(x,y) + \tfrac{1}{4}t\omega_2(x,p)\omega_2(y,p) + o(t) & x,y \in C_2-U_2 \end{cases} \tag{49}$$

where $\omega_1(x,y)$ and $\omega_2(x,y)$ are the bilinear differentials on C_1 and C_2, and the $o(t)$ are bilinear holomorphic differentials for which $\lim\limits_{t\to 0} \frac{1}{t} o(t) \equiv 0$ in x and y.

<u>Proof</u>. The expansion (48) comes from (47) of Prop. 3.1 and (30). Now if $\Omega(x,y,t) = \omega_t(x,y) - \omega_1(x,y) - \tfrac{1}{4}t\omega_1(x,p)\omega_1(y,p)$, then $\frac{1}{t^2}\Omega(x,y,t)$ is a holomorphic bilinear differential for $t \neq 0$ and $x,y \in C_1-U_1$ and, by (30) and (47), has bounded periods as $t \to 0$, so that actually $\lim\limits_{t\to 0} \frac{1}{t}\Omega(x,y,t) \equiv 0$ for all $x,y \in C_1-U_1$. The remaining expansions are similarly proved.

Let $\mathcal{J}$ be the quotient of $\mathbb{C}^g \times D$ under the identification of $(z,t) \in \mathbb{C}^g \times D$ with $(z + \kappa\tau(t) + 2\pi i\lambda\, ,\, t)$ for any $\kappa,\lambda \in \mathbb{Z}^g$; then $\mathcal{J} \overset{\pi}{\to} D$ is an analytic family of g-dimensional varieties over D for which the fibers $\mathcal{J}_t = \pi^{-1}(t) = J(C_t)$ for $t \neq 0$ and $\mathcal{J}_0 = \pi^{-1}(0) = J(C_0) = J(C_1) \times J(C_2)$. Likewise, choosing any fixed holomorphic section $p(t)$ of $\mathcal{C} \to D$ - for instance, a fixed point in C_1-U_1 or C_2-U_2 -

the spaces $J_n(C_t)$ for any $n \in \mathbb{Z}$ form an analytic family $\mathcal{J}_n \to D$ isomorphic to $\mathcal{J}$ under the mapping $\mathcal{A}_t \in J_n(C_t) \to \mathcal{A}_t - np(t) \in J_0(C_t)$. Since $\Theta_\tau(z)$ is a holomorphic function on $\mathbb{C}^g \times \mathcal{H}_g$, we can form the subvariety $\vartheta \subset \mathcal{J}$ of codimension 1, which is an analytic family $\vartheta \xrightarrow{\pi} D$ with fibers $\vartheta_t = \pi^{-1}(t)$ the theta divisor $(\Theta_t) \subset J(C_t)$ for $t \neq 0$ and $\vartheta_0 = (\Theta_0) = J(C_1) \times (\Theta_2) \sqcup (\Theta_1) \times J(C_2)$ in $J(C_0)$. That part of the divisor of Θ_t near $J(C_1) \times (\Theta_2)$, say, is given by $\zeta - \Delta(t)$ for positive divisors $\zeta = \mathcal{A} + \mathcal{B}$ of degree $g-1$ on C_t with $\mathcal{A}$ positive of degree g_1 with support in $C_1 - U_1$, and $\mathcal{B}$ positive of degree g_2-1 with support in C_2-U_2; a similar statement holds for the portion of (Θ_t) near $(\Theta_1) \times J(C_2)$. This observation leads to:

<u>Proposition 3.3</u>. Let $\Delta(t)$, Δ_1 and Δ_2 be the Riemann divisor classes for C_t, C_1 and C_2 respectively, and set $\phi(t) = \zeta - \Delta(t) \in (\Theta_t)$ with $\zeta = \mathcal{A} + \mathcal{B}$ a positive divisor of degree $g-1$ on C_t, as above. Then $\lim_{t\to 0} \Delta(t) = \Delta_1 + \Delta_2 + p \in J_{g-1}(C_0)$ and $\lim_{t\to 0} \phi(t) = (e,f) \in J(C_0)$ where $e = \mathcal{A} - p - \Delta_1 \in J(C_1)$ and $f = \mathcal{B} - \Delta_2 \in (\Theta_2)$. The condition $\frac{\partial}{\partial t} \Theta_t(\phi(t))\Big|_{t=0} = 0$ implies two identities valid on any Riemann surface C of genus g : for any basepoint $b \in C$ and $\Delta = (g-1)b + k^b \in J_{g-1}(C)$ given by (13):

$$- \sum_{i=1}^{g} v_i(p) \frac{\partial \ln \Theta}{\partial z_i}(e) = \int_{gb}^{\mathcal{A}} \omega(p,y) - \frac{1}{2\pi i} \sum_{i=1}^{g} \int_{A_i} v_i(x) \int_b^x \omega(p,y)$$

for $e = \mathcal{A} - p - \Delta \in \mathbb{C}^g$, $\mathcal{A}$ positive of degree g on C; and

$$\frac{d}{dp} \ln \Theta(p-b-f) = \int_{(g-1)b}^{\mathcal{B}} \omega(p,y) - \frac{1}{2\pi i} \sum_{j=1}^{g} \int_{A_j} v_j(x) \int_b^x \omega(p,y)$$

for $f = \mathcal{B} - \Delta \in (\Theta)$ non-singular, $\mathcal{B}$ positive of degree $g-1$ on C.*

---

* See Lemma 2.7 (32).

<u>Proof.</u>* Let $k^{b_1} \in \mathbb{C}^{g_1}$ and $k^{b_2} \in \mathbb{C}^{g_2}$ be the vector of Riemann constants for C_1 and C_2 with basepoints $b_1 \in C_1 - U_1$ and $b_2 \in C_2 - U_2$ respectively. Then the vector $k_t^{b_1} \in \mathbb{C}^g$ of Riemann constants for C_t has, by Cor. 3.2, an expansion of the form

$$k_t^{b_1} = \left(k^{b_1} + g_2 \int_{b_1}^{b_2} u(x,t)\ ,\ k^{b_2} + (g_2-1) \int_{b_1}^{b_2} u(x,t)\right) + td + o(t) \tag{50}$$

where $d = (d_1,\ldots,d_g) \in \mathbb{C}^g$ has components d_j for $j > g_1$ given by

$$\begin{aligned} d_j = &-\frac{1}{8} v_j(p)\, v_j(p) + \frac{1}{8\pi i} v_j(p) \sum_{\mu \le g_1} \int_{A_\mu} v_\mu(x) \int_{b_1}^{x} \omega_1(p,y) \\ &+ \frac{1}{8\pi i} \sum_{\substack{\nu > g_1 \\ \nu \ne j}} \int_{A_\nu} \left[v_j(p)\, v_\nu(x) \int_{b_2}^{x} \omega_2(p,y) + v_\nu(p)\, \omega_2(p,x) \int_{b_2}^{x} v_j(y) \right] \\ = &\ \frac{1}{8\pi i} v_j(p) \left[\sum_{\mu \le g_1} \int_{A_\mu} v_\mu(x) \int_{b_1}^{x} \omega_1(p,y) + \sum_{\nu > g_1} \int_{A_\nu} v_\nu(x) \int_{b_2}^{x} \omega_2(p,y) \right] \\ &+ \frac{1}{4}\left(\frac{v_j'(p)}{2} + v_j(p) \frac{d}{dp} \ln E(p,b_2) \right) \end{aligned}$$

by (13) and Stokes' Theorem applied to $\displaystyle\sum_{\nu > g_1} \int_{A_\nu B_\nu A_\nu^{-1} B_\nu^{-1}} \omega_2(x,p) \int_{b_2}^{x} v_j \int_{b_2}^{x} \omega_2(p,y)$.

By (50) and Cor. 3.2 again,

$$\begin{aligned} \phi(t) = &\left(\int_{g_1 b_1}^{\mathcal{A}} v - k^{b_1} - \int_{b_1}^{b_2} u(x,t)\ ,\ \int_{(g_2-1)b_2}^{\mathcal{B}} v - k^{b_2} \right) \\ &+ \frac{t}{4}\left[v(p) \left(\int_{g_1 b_1}^{\mathcal{A}} \omega_1(p,y) + \int_{(g_2-1)b_2}^{\mathcal{B}} \omega_2(p,y) \right) - 4d \right] + o(t) \end{aligned}$$

---

* In genus 3, a similar analysis has been carried out by Poincaré in [24].

and therefore

$$0 = \Theta_t(\phi(t)) = \sum_{m \in \mathbb{Z}^{g_1+g_2}} \exp\left\{\tfrac{1}{2} m\, T(t)\, m + m\cdot\phi(t)\right\}$$

$$= \frac{t}{4}\left\{ \theta_1(e) \sum_{j>g_1} \left[v_j(p)\left(\int_{g_1 b_1}^{A} \omega_1(p,y) + \int_{(g_2-1)b_2}^{\mathcal{B}} \omega_2(p,y)\right) - 4d_j \right] \frac{\partial\theta_2}{\partial z_j}(f) + \right.$$

$$\left. + \sum_{\substack{i\le g_1\\ j>g_1}} v_i(p)\, v_j(p) \frac{\partial\theta_1}{\partial z_i}(e)\frac{\partial\theta_2}{\partial z_j}(f) + \frac{1}{2}\theta_1(e)\sum_{j>g_1} v_{j_1}(p)\, v_{j_2}(p) \frac{\partial^2\theta_2}{\partial z_{j_1}\partial z_{j_2}}(f) \right\} + o(t)$$

where

$$(e,f) = \lim_{t\to 0}\phi(t) = \left(\int_{g_1 b_1}^{A} v - k^{b_1} - \lim_{t\to 0}\int_{b_1}^{b_2} u(x,t),\ \int_{(g_2-1)b_2}^{\mathcal{B}} v - k^{b_2} \right) \in J(C_1)\times(\Theta_2).$$

From the expression for d_j and the fact that

$$\frac{\displaystyle\sum_{j>g_1} v_{j_1}(p)v_{j_2}(p)\frac{\partial^2\theta_2}{\partial z_{j_1}\partial z_{j_2}}(f) - \sum_{j>g_1} v_j'(p)\frac{\partial\theta_2}{\partial z_j}(f)}{\displaystyle 2\sum_{j>g_1} v_j(p)\frac{\partial\theta_2}{\partial z_j}(f)} = \frac{d}{dp}\ln\frac{E(p,b_2)}{\theta_2(p-b_2-f)}$$

by Prop. 2.2, we conclude

$$-\sum_{i\le g_1} v_i(p)\frac{\partial \ln\theta_1}{\partial z_i}(e) - \int_{g_1 b_1}^{A}\omega_1(p,y) + \frac{1}{2\pi i}\sum_{i\le g_1}\int_{A_i} v_i(x)\int_{b_1}^{x}\omega_1(p,y)$$

$$= -\frac{d}{dp}\ln\theta_2(p-b_2-f) + \int_{(g_2-1)b_2}^{\mathcal{B}}\omega_2(p,y) - \frac{1}{2\pi i}\sum_{j>g_1}\int_{A_j} v_j(x)\int_{b_2}^{x}\omega_2(p,y),$$

and both sides of this equation vanish identically (consider the case $g_1 = 0$). Since $e \in (\Theta_1)$ whenever $p \in A$, $e = A - p - \Delta_1 \in J(C_1)$

and $\lim_{t\to 0}\int_{b_1}^{b_2} u_i(x,t) = \int_{b_1}^{p} v_i(x)$ for $i \le g_1$, which implies

$\lim_{t\to 0}\Delta(t) = \Delta_1 + \Delta_2 + p$ by (14) and (50).

Let us consider some further applications of the variational formulas (47)-(49).

Example 3.4. A generalization of Poincaré's asymptotic period relation in genus 4 [24, p. 291] is found by repeated use of the expansions of Cor. 3.2: Let $C_{t_1,\dots,t_g}$ be a Riemann surface of genus g obtained by attaching g torii $C_1,\dots,C_g$ punctured at g points $p_1,\dots,p_g$ to $\mathbb{P}_1(\mathbb{C})$ punctured at $q_1,\dots,q_g$ with pinching parameters $t_1,\dots,t_g \in D$ respectively. Then the Riemann matrix $\tau(t_1,\dots,t_g)$ for $C_{t_1,\dots,t_g}$ has the off-diagonal expansion for $i \neq j$:

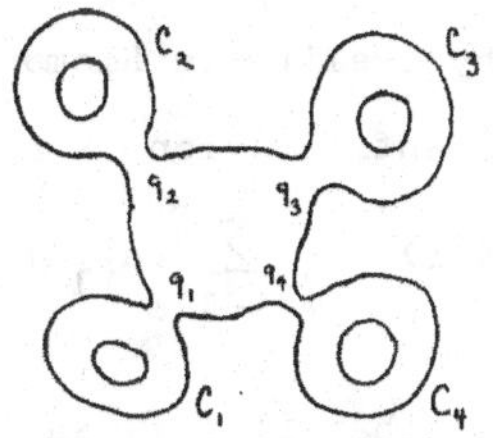

$$\tau_{ij}(t_1,\dots,t_g) = \frac{1}{16} t_i t_j v_i(p_i) v_j(p_j)\, \omega(q_i,q_j) + \text{higher order terms in } t_1,\dots,t_g$$

where $\omega(x,y) = \dfrac{dx\,dy}{(x-y)^2}$ is the differential of the second kind on $\mathbb{P}_1(\mathbb{C})$ with a double pole at $x = y$. Thus, neglecting terms of third-order or higher, the expressions $\dfrac{1}{\sqrt{\tau_{ij}}} = \dfrac{4(q_i-q_j)}{(t_i t_j v_i(p_i) v_j(p_j))^{\frac{1}{2}}}$ for $g \geq 4$ satisfy the asymptotic relations

$$\frac{1}{\sqrt{\tau_{ij}\tau_{k\ell}}} + \frac{1}{\sqrt{\tau_{ik}\tau_{\ell j}}} + \frac{1}{\sqrt{\tau_{i\ell}\tau_{jk}}} = 0$$

for all *distinct* $i,j,k,\ell \in 1,2,\dots,g$.

Example 3.5. When $g_2 = 0$, a case which has been extensively studied [25], the tangent $\left(v_1^2(p), v_1(p)v_2(p),\dots,v_g^2(p)\right) \in \mathbb{C}^{\frac{1}{2}g(g+1)}$ to a curve of period matrices in $\mathfrak{H}_g$ obtained by attaching a sphere at a point p in a non-hyperelliptic Riemann surface C will always lie in a 3g-3 dimensional subspace for any $p \in C$, since $v_1v_1, v_1v_2,\dots,v_gv_g$ span the space of quadratic differentials $H^0(\Omega_C^{\otimes 2}) \simeq \mathbb{C}^{3g-3}$. The rank of the $\frac{g(g+1)}{2} \times (3g-3)$ matrix $\left(v_i v_j(p_k)\right)$ is 3g-3 for generic points

$p_1,\ldots,p_{3g-3} \in C$, and we can obtain all period matrices of Riemann surfaces near τ by attaching spheres to C at points $p_1,\ldots,p_{3g-3}$ with pinching parameters $t_1,\ldots,t_{3g-3}$. The cotangent space to the locus $\mathcal{T}_g \subset \mathfrak{H}_g$ of period matrices of Riemann surfaces at a non-hyperelliptic Riemann surface C is thus isomorphic to $H^0(\Omega_C^{\otimes 2})$ via the identification

$$(51) \qquad \sum_{1\le i\le j\le g} a_{ij} d\tau_{ij} \longleftrightarrow \sum_{1\le i\le j\le g} a_{ij} v_i v_j \in H^0(\Omega_C^{\otimes 2}), \quad a_{ij} \in \mathbb{C}$$

so that if f is any function on $\mathfrak{H}_g$ vanishing on $\mathcal{T}_g$, the differential $df = \sum_{i\le j} \frac{\partial f}{\partial \tau_{ij}} d\tau_{ij}$ on $\mathfrak{H}_g$ vanishes identically when restricted to $\mathcal{T}_g$ - that is to say, the quadratic differential $\sum_{i\le j} \frac{\partial f}{\partial \tau_{ij}} v_i v_j$ vanishes identically on C.

It should be mentioned that the same variational formulas given in this chapter also arise from a family of curves obtained by varying branch points in a realization of the Riemann surface as a branched covering of the sphere; this approach has been worked out in [25]. The following result [31] illustrates the power of this method in the hyperelliptic case.

<u>Proposition 3.6</u> (Thomae). Let C be the hyperelliptic Riemann surface of genus g defined by $s^2 = \prod_1^{2g+2} (z-p_k)$, $p_k = z(Q_k) \in \mathbb{C}$, and suppose the normalized differentials $v_1,\ldots,v_g$ on C are given by $v_\alpha = \sum_1^g \sigma_{\alpha\beta} w_\beta$ where $w_\beta = \frac{z^{\beta-1}dz}{s}$ for $\beta = 1,\ldots,g$, and $(\sigma_{\alpha\beta})$ is a non-singular matrix depending on C. For any non-singular even half-period $e \in J_0(C)$, let $\{Q_{i_1},\ldots,Q_{i_{g+1}}\} \sqcup \{Q_{j_1},\ldots,Q_{j_{g+1}}\}$ be the partition of the Weierstrass points such that $e = \sum_1^{g+1} Q_{i_k} - D - \Delta$, as on p. 13. Then

$$\theta[e]^8(0) = (\det\sigma)^{-4} \prod_{\substack{k,\ell=1\\ k<\ell}}^{g+1} (p_{i_k}-p_{i_\ell})^2(p_{j_k}-p_{j_\ell})^2.$$

Proof. For any Weierstrass point $q \in C$ with $z(q) = p \in \mathbb{C}$, let $\mathcal{C} \to D$ be the family of hyperelliptic Riemann surfaces C_t over a t-disc defined by the equations $s^2(t) = (z-p-\frac{t}{2}) \prod_{\substack{1\\ p_k\neq p}}^{2g+2} (z-p_k)$ for each $t \in D$. By expanding the differentials $z^{\beta-1}dz/s(t)$ in a Taylor series at $t = 0$ as in the proof of Prop. 3.1, it can be seen that the normalized differentials and period matrix on the curves C_t satisfy exactly the same variational formulas as those given in (47)-(49). Now for any fixed $e \in \mathbb{C}^g$ with $\theta(e) \neq 0$, let $\mathcal{A} = \mathcal{A}(0) = \sum_1^g a_i = \operatorname{div}_C\theta(x-q-e)$ and, in general, $\mathcal{A}(t) = \sum_1^g a_i(t) = \operatorname{div}_{C_t}\theta_t(x-q(t)-e)$ where $q(t)$, for $t \in D$, is the Weierstrass point in C_t with $z(q(t)) = p+\frac{t}{2} \in \mathbb{C}$. Then since θ satisfies the heat equation: $\forall\, z \in \mathbb{C}^g$

(52) $$\frac{\partial^2\theta[e]}{\partial z_i\partial z_j}(z) = \frac{\partial\theta[e]}{\partial\tau_{ij}}(z),\quad i\neq j \quad\text{and}\quad \frac{\partial^2\theta[e]}{\partial z_i\partial z_i}(z) = 2\,\frac{\partial\theta[e]}{\partial\tau_{ii}}(z),$$

(35) implies that for *any* family of Riemann surfaces satisfying the variational formulas (47)-(49):

$$\frac{d}{dt}\ln\theta_t[e](0)\Big|_{t=0} = \sum_{1\le i\le j\le g}\frac{\partial\ln\theta[e]}{\partial\tau_{ij}}(0)\,\frac{d\tau_{ij}}{dt}\Big|_{t=0} = \frac{1}{8}\sum_{i,j=1}^{g}\frac{\partial^2\ln\theta[e]}{\partial z_i\partial z_j}(0)\,v_i(q)v_j(q)$$

$$= -\frac{1}{8}\sum_{i,j=1}^{g}\omega(q,a_i)\,v(\mathcal{A})^{-1}_{ij}\,v_j(q) = \frac{1}{2}\sum_{i,j=1}^{g} v(\mathcal{A})^{-1}_{ij}\left(-\frac{dv_{j,t}(a_i(t))}{dt} + v_j'(a_i)\,\frac{da_i(t)}{dt}\right)\Bigg|_{t=0}$$

If we specialize e to a non-singular even half-period and let q be, say, the point $Q_{i_{g+1}}$ in the partition corresponding to e, then $\sum_1^g a_k = \sum_1^g Q_{i_k}$, $v_j'(a_i) = 0 \;\forall\, i,j$ and the equation reduces to

$\frac{d}{dt} \ln \Theta_t[e](0)\sqrt{\det v_t(\mathcal{A}(t))}\Big|_{t=0} = 0$; since $\det v_t(\mathcal{A}(t)) =$

$\det \sigma(t) \det\limits_{1\le j,k\le g} \left(\frac{z(Q_{i_k})^{j-1}}{s_t(Q_{i_k})} \right)$, where $\sigma(t)$ is the matrix σ for C_t, this

condition becomes

$$0 = \frac{d}{dt} \ln \Theta_t^8[e](0) \frac{\det \sigma(t)^4}{\prod_1^g s_t(Q_{i_k})^4}\Bigg|_{t=0} = \frac{\partial}{\partial p_{i_{g+1}}} \ln \frac{\Theta^8[e](0)\,(\det\sigma)^4}{\prod_{k=1}^{g}(p_{i_k}-p_{i_{g+1}})^2}$$

A similar equation holds for $p_{j_{g+1}}$ and $p_{j_1},\dots,p_{j_g}$ since $e = \sum_1^{g+1} Q_{j_k} - D - \Delta$ also; interchanging $p_{i_{g+1}}$ (resp. $p_{j_{g+1}}$) with any $p_{i_1},\dots,p_{i_g}$ (resp. $p_{j_1},\dots,p_{j_g}$) we conclude that the function $\mathcal{U} =$ $\Theta^8[e](0)(\det\sigma)^4 \prod\limits_{1\le k<\ell\le g+1} (p_{i_k}-p_{i_\ell})^{-2}(p_{j_k}-p_{j_\ell})^{-2}$ has zero derivatives with respect to all branch points and hence is a universal constant on the moduli space of hyperelliptic Riemann surfaces. To evaluate $\mathcal{U}$, let C_ε be the hyperelliptic curve $s^2 = \prod\limits_{k=1}^{g+1} (z-p_{2k-1})(z-p_{2k-1}-\varepsilon)$ and C_0 the rational curve $s = \prod\limits_1^{g+1} (z-p_{2k-1})$: then as $\varepsilon \to 0$, the holomorphic differentials $v_k(\varepsilon)$, $k = 1,\dots,g$, normalized on C_ε with respect to the homology basis on p. 14, become differentials of the third kind $\left(\frac{dz}{z-p_{2k+1}} - \frac{dz}{z-p_1}\right)$ on C_0; and $\lim\limits_{\varepsilon\to 0} \Theta_{\tau(\varepsilon)}(0) = 1$ since the period matrix $\tau(\varepsilon)$ for C_ε has an expansion $(\ln\varepsilon)I + O(\varepsilon)$ with I the identity $g\times g$ matrix and $\lim\limits_{\varepsilon\to 0} O(\varepsilon)$ a finite matrix - see p. 53. Also, for $\mu = 1,\dots,g$ and δ sufficiently small,

$$\lim_{\varepsilon\to 0} \int_{A_\mu} w_\beta(\varepsilon) = \int_{|z-p_{2\mu+1}|=\delta} \frac{z^{\beta-1}\,dz}{\prod_1^{g+1}(z-p_{2k-1})} = \frac{2\pi i\; p_{2\mu+1}^{\beta-1}}{\prod\limits_{\substack{k=1\\ k\ne\mu+1}}^{g+1} (p_{2\mu+1}-p_{2k-1})}$$

so that if $\sigma(\varepsilon)$ is the σ matrix for C_ε,

$$(2\pi i)^g = \lim_{\epsilon\to 0} \det\left(\int_{A_\mu} v_\alpha(\epsilon)\right) = \lim_{\epsilon\to 0} \det\sigma(\epsilon) \frac{(2\pi i)^g \det(p_{2\mu+1}^{\beta-1})}{\prod_{\mu=1}^{g} \prod_{\substack{k=1 \\ k\neq\mu+1}}^{g+1} (p_{2\mu+1} - p_{2k-1})}$$

and thus

$$\lim_{\epsilon\to 0} \det\sigma(\epsilon)^2 = \det(p_{2\ell+1}^{\beta-1})^{-2} \prod_{\substack{0\le k,\mu\le g \\ \mu\neq 0,k}} (p_{2\mu+1} - p_{2k+1})^2 = \prod_{0\le k<\mu\le g} (p_{2\mu+1} - p_{2k+1})^2$$

However, from the computation (18) of the class Δ, the half-period $e = 0$ corresponds to the partition $\{Q_1, Q_3, \ldots, Q_{2g+1}\} \cup \{Q_2, Q_4, \ldots, Q_{2g+2}\}$; so we finally can conclude that

$$U = \lim_{\epsilon\to 0} \frac{\theta_{T(\epsilon)}^{8}(0)}{\det\sigma(\epsilon)^{-4}} \prod_{1\le k<\ell\le g+1} (p_{2k-1} - p_{2\ell-1})^{-2} (p_{2k-1} + \epsilon - p_{2\ell-1} - \epsilon)^{-2} = 1 .$$

As a corollary, we see that for any two half-periods e and e' corresponding to partitions $\{i_1, \ldots, i_{g+1}\} \cup \{j_1, \ldots, j_{g+1}\}$ and $\{i_1', \ldots, i_{g+1}'\} \cup \{j_1', \ldots, j_{g+1}'\}$ respectively,

$$\frac{\theta[e]^4(0)}{\theta[e']^4(0)} = \pm \prod_{k,\ell=1}^{g+1} \frac{(p_{i_k} - p_{i_\ell})(p_{j_k} - p_{j_\ell})}{(p_{i_k'} - p_{i_\ell'})(p_{j_k'} - p_{j_\ell'})}$$

which gives, in the elliptic case (p. 34),

$$\frac{\theta\left[\begin{smallmatrix}1/2\\0\end{smallmatrix}\right]^4(0)}{\theta\left[\begin{smallmatrix}0\\0\end{smallmatrix}\right]^4(0)} = \frac{\wp(\pi i + \frac{T}{2}) - \wp(\frac{T}{2})}{\wp(\pi i) - \wp(\frac{T}{2})} = \lambda\left(\frac{T}{2\pi i}\right) .$$

These classical formulas can be proved directly from the factorization of rational functions on hyperelliptic Riemann surfaces:

$$\frac{\theta(\sum_1^g x_i - a - \Delta)\, \theta(\sum_1^g x_i - \phi(a) - \Delta)}{\theta(\sum_1^g x_i - b - \Delta)\, \theta(\sum_1^g x_i - \phi(b) - \Delta)} = c_{a,b} \prod_{i=1}^{g} \frac{z(x_i) - z(a)}{z(x_i) - z(b)} \qquad \forall\, a,b \in C$$

where ϕ is the involution on C and $c_{a,b}$ is a constant independent of $x_1,\ldots,x_g \in C$; see, for instance, [31].

Pinching a Non-Zero Homology Cycle. Here $\mathcal{C} \to D$ will be a family of Riemann surfaces over the unit t-disc D constructed in the following manner: let C be a compact Riemann surface of genus g and choose coordinates $z_a\colon U_a \xrightarrow{\sim} D$ and $z_b\colon U_b \xrightarrow{\sim} D$ in disjoint neighborhoods U_a and U_b of two points $a,b \in C$. Set

$$W = \{(p,t) \mid t \in D,\ \ p \in C\text{-}U_a\text{-}U_b \ \text{ or } \ p \in U_a \ \ (\text{resp. } U_b)$$
$$\text{with } |z_a(p)| > |t| \ \ (\text{resp. } |z_b(p)| > |t|)\}$$

and let S be the surface $\{XY = t \mid (X,Y,t) \in D \times D \times D\}$. Then define $\mathcal{C} = W \sqcup S$ where, in the overlap,

$$(p_a,t) \in W \cap U_a \times D \ \text{ is identified with } \ (z_a(p_a), \frac{t}{z_a(p_a)}, t) \in S$$

and

$$(p_b,t) \in W \cap U_b \times D \ \text{ is identified with } \ (\frac{t}{z_b(p_b)}, z_b(p_b), t) \in S.$$

Again $x = \frac{1}{2}(X+Y)$ and $y = \frac{1}{2}(X-Y)$ will be coordinates on S so that each fiber C_t of $\mathcal{C}$ for $t \neq 0$ is a Riemann surface of genus $g+1$ for which the pinched region $C_t \cap S$ is a ramified double covering $y = \sqrt{x^2-t}$ of a neighborhood of $x = 0$ with branch points at $x = \pm\sqrt{t}$. The fiber C_0 is a curve of genus g with an ordinary double point corresponding to the points a,b in the normalization C of C_0; the branches of $C_0 \cap S$ corresponding to neighborhoods of $a,b \in C$ are $y = x$, $y = -x$ with local pinching coordinates $x = \frac{1}{2}z_a$ and $x = \frac{1}{2}z_b$, respectively.

To choose some canonical homology basis for C_t, let $A_1(t),B_1(t),\ldots,A_g(t),B_g(t)$ simply be a canonical base $A_1,B_1,\ldots,A_g,B_g$ for C lying in $C\text{-}U_a\text{-}U_b$ extended across $(C\text{-}U_a\text{-}U_b) \times D$. Set $A_{g+1}(t) = \partial U_b \times \{t\}$ and, for $|t| < \frac{1}{4}$, $B_{g+1}(t) = \gamma \times \{t\} \sqcup \gamma_{at} \sqcup \gamma_{bt} \subset W$ where

γ is any fixed path from $z_a^{-1}(\tfrac{1}{2})$ to $z_b^{-1}(\tfrac{1}{2})$ lying within C cut along its homology basis, and γ_{at} and γ_{bt} are continuously varying paths from $z_a^{-1}(\sqrt{t})$ to $z_a^{-1}(\tfrac{1}{2})$ and from $z_b^{-1}(\tfrac{1}{2})$ to $z_b^{-1}(\sqrt{t})$ lying in $|\sqrt{t}| < |z_a| < 1$ and $|\sqrt{t}| < |z_b| < 1$ respectively, so that $B_{g+1}(0)$ is a path from a to b in C. As t goes once around the origin, any fixed continuous determination of $B_{g+1}(t)$ increases by a cycle homologous to $\pm A_{g+1}(t)$, and thus a well-defined choice of $B_{g+1}(t)$ can be given only in the t-disc cut along some path from the origin.

<u>Proposition 3.7.</u> There are $g+1$ linearly independent holomorphic 2-forms on $\mathcal{C}$ whose residues $u_1(x,t),\ldots,u_{g+1}(x,t)$ along C_t for t in a sufficiently small disc D_ε of radius ε about $t = 0$ are a normalized basis for the holomorphic differentials on C_t for $t \neq 0$; while, for $t = 0$, the differentials $u_i(x,0)$, $i = 1,\ldots,g$ are a normalized basis for the differentials on C and $u_{g+1}(x,0)$ is $\omega_{b-a}(x)$, the normalized differential of the third kind on C with simple poles of residue $-1,+1$ at a,b. For $x \in C-U_a-U_b$ and $i = 1,2,\ldots,g$,

(53)
$$u_i(x,t) = v_i(x) + \tfrac{1}{4}t(v_i(a) - v_i(b))(\omega(x,a) - \omega(x,b)) + O(t^2),$$
and
$$u_{g+1}(x,t) = \omega_{b-a}(x) + t\tilde{u}_{g+1}(x) + O(t^2)$$

where $v_1(x),\ldots,v_g(x)$ are the normalized differentials on C, $\omega(x,y)$ is the differential of the second kind on C, the expressions $O(t^2)$ are holomorphic differentials on $C-U_a-U_b$ with $\lim_{t\to 0} \frac{1}{t^2}O(t^2)$ a finite differential there, and $\tilde{u}_{g+1}(x)$ is a normalized differential of the second kind on C with only poles of order 3 at a and b, where the Laurent expansions are, in terms of the pinching coordinates:

$$\pm\left(-\frac{1}{2x^3} + \frac{\beta}{x^2} + \ldots\right)dx \quad \text{with} \quad 4\beta = \frac{d}{db}\ln E(b,a) + \frac{d}{da}\ln E(a,b).$$

<u>Proof.</u> If $\Omega^2_{\mathcal{C}}$ is the sheaf of holomorphic 2-forms on $\mathcal{C}$, then under the residue map, $\Omega^2_{\mathcal{C}}|C_t$ is the sheaf of holomorphic differentials

on C with simple poles of opposite residue at a and b for $t = 0$; so $\dim H^0(C_t, \Omega^2_{\mathcal{C}} | C_t) = g+1 \ \forall t$ and, as in Prop. 3.1, Grauert's Theorem again implies that $\pi_* \Omega^2_{\mathcal{C}}$ is a locally free sheaf on D of rank $g+1$. Thus for t near 0, there are holomorphic forms $U_1(x,t), \ldots, U_g(x,t), U_{g+1}(x,t)$ on $\mathcal{C}$ whose residues along $t = 0$ give, respectively, a normalized basis $v_1(x), \ldots, v_g(x)$ of the differentials on C and $\omega_{b-a}(x)$, the normalized differential of the third kind with poles at a,b. The holomorphic matrix $\frac{1}{2\pi i}\left(\int_{A_j(t)} \operatorname{Res}_{C_t} U_i(x,t)\right)$ is the identity matrix at $t = 0$ and is invertible near $t = 0$; by changing the basis $\{U_i,\ i = 1, \ldots, g+1\}$ by this matrix and taking the residues along C_t we then get a normalized basis $u_1(x,t), \ldots, u_{g+1}(x,t)$ for the differentials on C_t. Now for $i < g+1$ and $x \in C_t \cap S$ in a neighborhood of the double point, let

$$u_i(x,t) = \sum_0^\infty a_\mu(t) x^\mu dx + \sum_0^\infty b_\nu(t) \frac{x^\nu}{\sqrt{x^2-t}} dx$$

in the pinching coordinate x, with a_μ and b_ν holomorphic functions near $t = 0$. Then

$$u_i(x,0) = \sum_0^\infty a_\mu(0) x^\mu dx \pm \sum_0^\infty b_\nu(0) x^{\nu-1} dx = v_i(x)$$

so that $b_0(0) = 0$, $v_i(a) = a_0(0) + b_1(0)$, $v_i(b) = a_0(0) - b_1(0)$ and

$$\lim_{t\to 0} \frac{u_i(x,t) - v_i(x)}{t} = \sum_{\mu\geq 0} a'_\mu(0) x^\mu dx \pm \sum_{\nu \geq 0}\left(\tfrac{1}{2} b_{\nu+1}(0) + x b'_\nu(0)\right) x^{\nu-2} dx$$

is a normalized differential of the second kind on C with only double poles of zero residue at a and b where the Laurent expansions have leading coefficients $\pm\frac{1}{2} b_1(0) = \pm\frac{1}{2}(v_i(a) - v_i(b))$ in terms of the pinching coordinates. On the other hand, if

$$u_{g+1}(x,t) = \sum_0^\infty \alpha_\mu(t) x^\mu dx + \sum_0^\infty \beta_\nu(t) \frac{x^\nu}{\sqrt{x^2-t}} dx \qquad x \in C_t \cap S$$

for holomorphic α_μ, β_ν, then $\beta_0(0) = -1$ since $u_{g+1}(x,0) = \omega_{b-a}(x)$. Thus

$$\lim_{t\to 0} \frac{u_{g+1}(x,t)-\omega_{b-a}(x)}{t} = \sum_0^\infty a_\mu'(0)x^\mu dx \pm \sum_0^\infty (\tfrac{1}{2}\beta_\nu(0)x^{\nu-3} + \beta_\nu'(0)x^{\nu-1})dx$$

is a normalized differential of the second kind with only triple poles at a and b where the Laurent developments begin $\pm(-\frac{1}{2x^3} + \frac{\beta}{x^2} + \text{holom.})dx$ with, from (21):

$$4\beta = 2\beta_1(0) = \lim_{\delta\to 0}\left(-\omega_{b-a}(b+\delta) + \omega_{b-a}(a+\delta) + \frac{2}{\delta}\right) = \frac{d}{db}\ln E(b,a) + \frac{d}{da}\ln E(a,b)$$

in terms of the pinching coordinates.

<u>Corollary 3.8</u>. The Riemann matrix for C_t has an expansion

$$(54) \qquad T(t) = \left(\begin{array}{c|c} \tau_{ij} + t\sigma_{ij} & a_i + t\sigma_{ig} \\ \hline a_j + t\sigma_{gj} & \ln t + c_1 + c_2 t \end{array}\right)_{1\le i,j\le g} + O(t^2)$$

for some constants c_1, c_2, where (τ_{ij}) is the Riemann matrix for C, $a_i = \int_a^b v_i$, $\sigma_{ij} = \tfrac{1}{4}(v_i(a)-v_i(b))(v_j(a)-v_j(b))$, $\sigma_{ig} = \sigma_{gi} = \tfrac{1}{4}(v_i'(b)-v_i'(a)) + \beta(v_i(a)-v_i(b))$ and $\lim_{t\to 0}\frac{1}{t^2}O(t^2)$ is a finite matrix. The differential of the second kind on C_t has an expansion for all $x,y \in C-U_a-U_b$:

$$\omega_t(x,y) = \omega(x,y) + \frac{t}{4}(\omega(x,a)-\omega(x,b))(\omega(y,a)-\omega(y,b)) + O(t^2)$$

with $\omega(x,y)$ the bilinear differential for C and $\lim_{t\to 0}\frac{1}{t^2}O(t^2)$ a meromorphic differential on C.

<u>Proof</u>. Prop. 3.7 and the general bilinear relation for differentials of the second kind [14, p. 176] give everything except the entry $\tau_{g+1,g+1}(t)$. But from the statement preceding Prop. 3.7, $\tau_{g+1,g+1} - \ln t$ is a well-defined analytic function of t in the

punctured disc $D_\varepsilon - \{0\}$, which must actually be analytic in the entire disc D_ε since otherwise $\operatorname{Re} \tau(t)$ would not be negative definite as $t \to 0$.

As an example, let $C_{t_1,\ldots,t_g}$ be a Riemann surface of genus g which is being pinched along $A_1,\ldots,A_g$ with parameters $t_1,\ldots,t_g \in D$ so that $C_{0,0,\ldots,0}$ is of genus 0 with g double points corresponding to g pairs of points $a_1,b_1,\ldots,a_g,b_g \in \mathbb{P}_1(\mathbb{C})$. The Riemann matrix for $C_{t_1,\ldots,t_g}$ has an expansion

$$\left.\begin{aligned} \tau_{ii}(t_1,\ldots,t_g) &= \ln t_i + \text{constant} \\ \tau_{ij}(t_1,\ldots,t_g) &= (a_i,b_i;a_j,b_j) \end{aligned}\right\} + \text{higher order terms in } t_1,\ldots,t_g$$

where (;) is the cross ratio of four points in $\mathbb{P}_1$.

From Cor. 3.8 we see that two points in $\mathbb{C}^{g+1}$ which differ by a point in the lattice Γ_t of rank 2g+2 generated by the columns of the matrix $(2\pi i I, \tau(t))$ must differ, as $t \to 0$, by a point in the lattice Γ_0 of rank 2g+1 generated by the columns of the matrix

$$\left(2\pi i I,\ \begin{matrix} \tau_{11}, & \cdots, & \tau_{1g} & 0 \\ \vdots & & \vdots & \vdots \\ \tau_{g1}, & \cdots, & \tau_{gg} & 0 \\ \int_a^b v_1, & \cdots, & \int_a^b v_g & 0 \end{matrix}\right)$$

We let, therefore, $\mathcal{J} \xrightarrow{\pi} D$ be the family of (g+1)-dimensional manifolds over D with fiber $\pi^{-1}(t)$ for $t \neq 0$ given by the Jacobian variety $J(C_t) = \mathbb{C}^{g+1}/\Gamma_t$, and with $\pi^{-1}(0)$ the non-compact Abelian group $\mathcal{J}_0 = \mathbb{C}^{g+1}/\Gamma_0$; then it can be shown that $\mathcal{J}$ has the structure of a complex manifold and that the projection $\mathcal{J} \xrightarrow{\pi} D$ is an analytic mapping - see the lecture by Jambois in [35, p. 30]. To describe the fiber $\mathcal{J}_0$, observe that if p is the double point on C_0, any divisor $\mathcal{A} = \sum_{a_i \neq p} n_i a_i + np$ on C_0 can be lifted to a divisor $\mathcal{A} = \sum_{a_i \neq p} n_i a_i + n(a+b)$ on the Riemann surface C; so if $\mathcal{A}$ and $\mathcal{B}$ are any two divisors of the same degree on C_0 such that $\mathcal{B} - \mathcal{A}$ is the divisor of a meromorphic function f on C_0 lifted to C, Abel's

Theorem (8) gives

$$\frac{f(b)}{f(a)} = \exp\left\{\int_{\mathcal{A}}^{\mathcal{B}} \omega_{b-a} - \sum_{1}^{g}\int_{a}^{b} m_i v_i\right\} \in ¢^*, \qquad \begin{Bmatrix} m \\ * \end{Bmatrix}_{\tau} = \int_{\mathcal{A}}^{\mathcal{B}} v \in ¢^g$$

which holds even if f has a zero or pole at a since it must also have the same zero or pole at b. Thus if we let the divisor $\mathcal{B} - \mathcal{A}$ of degree 0 on C_0 correspond to the equivalence class of $\left(\int_{\mathcal{A}}^{\mathcal{B}} v_1, \ldots, \int_{\mathcal{A}}^{\mathcal{B}} v_g, \int_{\mathcal{A}}^{\mathcal{B}} \omega_{b-a}\right) \in ¢^{g+1}$ modulo Γ_0, the variety $\mathcal{J}_0$ becomes the group of divisor classes of degree 0 on C_0 with two divisors D and D' identified if $D - D'$ is the divisor of a meromorphic function on C_0 - that is, a function f on C satisfying $f(b)/f(a) = 1$. There is an exact sequence of groups

$$(55) \qquad 0 \longrightarrow ¢^* \xrightarrow{\ \phi\ } \mathcal{J}_0 \xrightarrow{\ \psi\ } J_0(C) \longrightarrow 0$$

where ψ is induced from the identity on divisors of C_0 lifted to C, and $\phi(r)$ for $r \in ¢^*$ is the class in $\mathcal{J}_0$ of the divisor of any meromorphic function f on C satisfying $\frac{f(b)}{f(a)} = r$. We will let $Z = (z_1, \ldots, z_g, z_{g+1}) \in ¢^{g+1}$ denote a point in the universal cover of $\mathcal{J}_0$ so that with this notation, $\phi(r)$ is the class of $(0, \ldots, 0, \ln r)$ modulo Γ_0 and $\psi(Z)$ is the class of $z = (z_1, \ldots, z_g) \in ¢^g$ in $J_0(C)$.

<u>Proposition 3.9</u>. Let $\delta(t)$ be the half-period $\frac{1}{2}\tau_{g+1}(t) = \begin{Bmatrix} 0 & \ldots & 0 & \frac{1}{2} \\ 0 & \ldots & 0 & 0 \end{Bmatrix}_{\tau(t)} \in ¢^{g+1}$. Then there is an analytic subvariety $\vartheta_\delta \subset \mathcal{J}_0$ of codimension 1 which is a family $\vartheta_\delta \to D$ of g-dimensional varieties over D with fibers at $t \neq 0$ given by $\operatorname{div}_{J(C_t)} \theta_t(Z - \delta(t))$, while for $t = 0$ the fiber is the subvariety of $\mathcal{J}_0$ defined by

$$(56) \qquad e^{z_{g+1}} + \frac{\theta\left(z - \frac{1}{2}\int_a^b v\right)}{\theta\left(z + \frac{1}{2}\int_a^b v\right)} = 0, \qquad z_{g+1} \in ¢ \ \text{ and } \ z \in ¢^g$$

where θ is the theta function for C.

Proof. The eigenvalues of the Riemann matrix τ of C are bounded away from 0 by $2\lambda < 0$, say; and thus the expansion (54) implies that $\sum_1^g n_i n_j \operatorname{Re} \tau_{ij}(t) < \lambda \sum_1^g n_i^2$ for t near 0 and $n_i \in \mathbb{R}$. By fixing $Z \in \mathbb{C}^{g+1}$ and expanding $\delta(t)$ by Cor. 3.8:

$$|\Theta_{\tau(t)}(Z - \delta(t))| \leq \sum_{m \in \mathbf{Z}} (|t| e^{\alpha(t)})^{\frac{1}{2}(m^2-m)} e^{mc} \prod_1^g \theta_\lambda(\beta_i(t) + m\gamma_i(t))$$

where $\theta_\lambda(w) = \sum_{n \in \mathbf{Z}} \exp(\frac{1}{2}n^2\lambda + nw)$, $c = \operatorname{Re} z_{g+1}$, and $\alpha(t)$, $\beta_i(t)$ and $\gamma_i(t)$ are the real parts of analytic functions bounded near $t = 0$. From this we conclude that for t sufficiently near 0, the above series converges by the ratio test and $\Theta_{\tau(t)}(Z - \delta(t))$ is a well-defined analytic function of Z and t for t near 0. The constant term in the Taylor development is

$$\lim_{t \to 0} \Theta_{\tau(t)}(Z - \delta(t)) = \theta(z - \tfrac{1}{2}\int_a^b v) + e^{z_{g+1}} \theta(z + \tfrac{1}{2}\int_a^b v)$$

which gives (56).

Thus, although the Riemann divisor class $\Delta(t) \in J_g(C_t)$ corresponding to (Θ_t) is not single valued as t goes once around the origin, $\Delta(t)+\delta(t)$ *is* a well-defined point in $J_g(C_t)$, and the bundle of half-order differentials L_α on C_t for any half-period $[\alpha] = \begin{bmatrix} \delta_1 & \cdots & \delta_g & \delta_{g+1} \\ \varepsilon_1 & \cdots & \varepsilon_g & \varepsilon_{g+1} \end{bmatrix}$ is likewise well-defined if and only if $\delta_{g+1} = \frac{1}{2}$. It will now be shown that

$$\psi_g[\lim_{t \to 0}(\Delta(t)+\delta(t))] = \Delta + \tfrac{1}{2}(a+b) \in J_g(C)$$

where Δ is the Riemann divisor class for C and, for any divisors $\mathcal{A}$ and $\mathcal{B}$ of degree g, $\psi_g(\mathcal{A}) = \mathcal{B} + \psi(\mathcal{A} - \mathcal{B}) \in J_g(C)$ with ψ the map in (55); here $\frac{1}{2}(a+b) \in J_1(C)$ is given by $c + \frac{1}{2}\int_{2c}^{a+b} v$ for any $c \in C$ with integration taken in C cut along its homology basis.

Proposition 3.10. Let $f(t) = \mathcal{A} - \delta(t) - \Delta(t) \in J_0(C_t)$ with $\mathcal{A}$ a positive divisor of degree g with support in $C - U_a - U_b$. Then

$$\psi(\lim_{t\to 0} f(t)) = \mathcal{A} - \frac{a+b}{2} - \Delta \in J_0(C)$$

and the condition $\lim_{t\to 0} \Theta_t(f(t)-\delta(t)) = 0$ becomes

$$\frac{\Theta(e - \frac{1}{2}\int_a^b v)}{\Theta(e + \frac{1}{2}\int_a^b v)} = \exp\left\{\int_{gq}^{\mathcal{A}} \omega_{b-a} - \frac{1}{2\pi i}\sum_{k=1}^{g}\int_{A_k} v_k(x)\int_q^x \omega_{b-a}\right\} \quad \forall\, q \in C,$$

where $e = \mathcal{A} - \frac{a+b}{2} - \Delta \in \mathbb{C}^g$ and Δ is given by (13)-(14).*

Proof. Let $(k_{it}^q) \in \mathbb{C}^{g+1}$ and $(k_{i0}^q) \in \mathbb{C}^g$ be the vectors of Riemann constants for C_t and C_0 with basepoint $q \in C - U_a - U_b$; then the expansions (53)-(54) give

$$k_{jt}^q - k_{j0}^q + \tfrac{1}{2}\tau_{j,g+1}(t) = \frac{1}{2\pi i}\int_{A_{g+1}} \omega_{b-a}(x)\int_q^x v_j - \tfrac{1}{2}\int_a^b v_j + O(t)$$

$$= \int_q^b v_j - \tfrac{1}{2}\int_a^b v_j + O(t) = \tfrac{1}{2}\int_{2q}^{a+b} v_j + O(t)$$

for $j \leq g$, and

$$(57)\quad k_{g+1,t}^q + \tfrac{1}{2}\tau_{g+1,g+1}(t) = \pi i + \frac{1}{2\pi i}\sum_{k=1}^{g}\int_{A_k} v_k(x)\int_q^x \omega_{b-a} + O(t).$$

Thus $\lim_{t\to 0} f(t)$ is the class in $\mathcal{J}_0$ of the point $Z = (z, z_{g+1}) \in \mathbb{C}^{g+1}$, where

$$z = \int_{gq}^{\mathcal{A}} v - k_0 - \tfrac{1}{2}\int_{2q}^{a+b} v_j = \mathcal{A} - \frac{a+b}{2} - \Delta \in \mathbb{C}^g$$

and

* See also Lemma 2.7 (33).

$$z_{g+1} = \int_{gq}^{\mathcal{A}} \omega_{b-a} - \pi i - \frac{1}{2\pi i} \sum_{1}^{g} \int_{A_k} v_k(x) \int_q^x \omega_{b-a} \in \mathbb{C}^g$$

The proposition now comes from applying (56) of Prop. 3.9.

The group $\mathcal{J}_0$ is the setting for a Jacobi inversion problem for differentials of the third kind:

Proposition 3.11. Let a and b be two distinct points on a Riemann surface C of genus g and suppose $\mathcal{A}$ is a positive divisor of degree $g+1$ not containing a or b. Then for any $r \in \mathbb{C}^*$ and $e \in \mathbb{C}^g$ with $\Theta(\mathcal{A}-a-b-\Delta+e) \neq 0$, there is a unique positive divisor $\mathcal{B}$ of degree $g+1$ not containing a or b such that

$$(\star) \qquad \left(\int_{\mathcal{A}}^{\mathcal{B}} v_1, \ldots, \int_{\mathcal{A}}^{\mathcal{B}} v_g, \int_{\mathcal{A}}^{\mathcal{B}} \omega_{b-a}\right) = (e_1, \ldots, e_g, \ln r) \in \mathbb{C}^{g+1} \quad \text{modulo } \Gamma_0.$$

For generic $\mathcal{A} = \sum_{1}^{g+1} a_i$, $\mathcal{B}$ is the divisor of zeroes of the section of $e \in J_0(C)$ given by

$$s(x) = \frac{r\Theta(x+\Delta+a-\mathcal{A}-e) - s(a,b,x,\mathcal{A})\Theta(x+\Delta+b-\mathcal{A}-e)}{\Theta(x+\Delta+a-\mathcal{A}) - s(a,b,x,\mathcal{A})\Theta(x+\Delta+b-\mathcal{A})}$$

with poles at $x \in \mathcal{A}$, where

$$s(a,b,x,\mathcal{A}) = \frac{E(b,x)}{E(a,x)} \prod_{1}^{g+1} \frac{E(a,a_i)}{E(b,a_i)} \exp\left\{\frac{1}{2\pi i} \sum_{k=1}^{g} \int_{A_k} v_k(y) \ln \frac{E(y,b)}{E(y,a)}\right\}$$

is a section of the bundle $b-a \in J_0$ with a simple zero and pole at b and a, respectively.

Proof. If $\Theta(\mathcal{A}-a-b-\Delta+e) \neq 0$, $\dim H^0(\mathcal{A}-a-b+e) = 0$ and $\dim H^0(\mathcal{A}-a-e) = \dim H^0(\mathcal{A}-b-e) = 1$, so we can find multiplicative meromorphic functions $s_1(x)$ and $s_2(x)$, sections of the flat bundle $e \in J_0$, with poles at $\mathcal{A}$ and zeroes at a and b respectively, and $s_1(b)s_2(a) \neq 0$. Then $s(x) = s_1(b)s_2(x) + rs_2(a)s_1(x)$ is a section of $e \in J_0$ with poles at $\mathcal{A}$ and $r = s(b)/s(a)$, and the divisor $\mathcal{B}$ of zeroes of $s(x)$ provides a solution to the inversion problem (★) for $r \in \mathbb{C}^*$, $e \in J_0$. This solution $\mathcal{B}$ is unique since if $\hat{\mathcal{B}}$ were a second solution to (★), $\hat{\mathcal{B}} - \mathcal{B}$ would be the divisor of a meromorphic function $f(x)$ on C with $f(b)/f(a) = 1$ so that $f(x) - f(a) \in H^0(\mathcal{B} - a - b)$ and $\dim H^0(\mathcal{A} - a - b + e) \geq 1$. The formula for $s(x)$ can be proved by checking that $s(x)$, a section of e satisfying $s(b)/s(a) = r$, has denominator vanishing for $x \in \mathcal{A}$ by repeated use of (34) of Lemma 2.7. The numerator of $s(x)$ can also be found by the classical Jacobi Inversion Theorem on the curves C_t for $t \neq 0$: if $\mathcal{B}(t) - \mathcal{A}$ is the class in $J(C_t)$ of the point $\hat{e} = (e_1, \ldots, e_g, \ln r) \in \mathbb{C}^{g+1}$, then $x \in \mathcal{B}(t)$ iff $x + \Delta(t) + \delta(t) - \mathcal{A} - \hat{e}$ is a point on the variety ϑ_δ of Prop. 3.9. Since

$$\lim_{t\to 0} \psi(x + \Delta(t) + \delta(t) - \mathcal{A} - \hat{e}) = x + \Delta + \frac{a+b}{2} - \mathcal{A} - e \in J_0(C)$$

by Prop. 3.10, and

$$\lim_{t\to 0} \int_{\mathcal{A}-\delta(t)}^{x+\Delta(t)} u_{g+1,t} - \ln r = -\ln r + \pi i + \int_{\mathcal{A}}^{x+gq} \omega_{b-a} + \frac{1}{2\pi i}\sum_{k=1}^{g} \int_{A_k} v_k(y) \int_q^y \omega_{b-a}$$

by (53) and (57), we conclude from Prop. 3.9 that

$$\mathcal{B} = \lim_{t\to 0} \mathcal{B}(t) = a + \operatorname{div}_C[\theta(x+\Delta+a-\mathcal{A}-e) + e^{z_{g+1}}\theta(x+\Delta+b-\mathcal{A}-e)]$$

where

$$\exp z_{g+1} = \exp\left\{\lim_{t\to 0} \int_{\mathcal{A}-\delta(t)}^{x+\Delta(t)} u_{g+1,t} - \ln r\right\} = -\frac{1}{r}\,\frac{E(b,x)}{E(a,x)} \prod_{1}^{g+1} \frac{E(a,a_i)}{E(b,a_i)} \exp\left\{\frac{1}{2\pi i}\sum_{k=1}^{g}\int_{A_k} v_k(y) \ln\frac{E(b,y)}{E(a,y)}\right\}.$$

In the exceptional case when $\dim H^0(\mathcal{A} - a - b + e) \geq 1$, that is to say $\theta(\mathcal{A} - a - b - \Delta + e) = 0$, the solution $\mathcal{B}$ to (★) must contain the points a and b unless further conditions are imposed on $r \in \mathbb{C}^*$. For instance, suppose a and b are two points on a torus C and $a_1, a_2 \in C$ are distinct from a,b; then

$$h(x) = \frac{r\sigma(x+a-a_1-a_2-e)\sigma(x-a) - k\sigma(x+b-a_1-a_2-e)\sigma(x-b)}{\sigma(x-a_1)\sigma(x-a_2)} \exp(-\eta e \int^x dz)$$

for $k = \sigma(a-a_1)\sigma(a-a_2) / \sigma(b-a_1)\sigma(b-a_2) \exp \eta e(b-a) \in ¢^*$, is a section of $e \in J_0$ with poles at a_1, a_2 and with $h(a)h(b) \neq 0$ and $h(b)/h(a) = r \in ¢^*$ if $\sigma(b+a-a_1-a_2-e) \neq 0$. On the other hand, if $\mathcal{B} = b_1+b_2$ is a solution to the inversion problem for $r \in ¢^*$, $\mathcal{A} = a_1+a_2$ and an exceptional $e \in ¢$ satisfying $\sigma(b+a-a_1-a_2-e) = 0$, then $\mathcal{B} - \mathcal{A} = \mathrm{div}_C h$ where

$$\text{(i)} \qquad h(x) = \lambda \frac{\sigma(x-b_1)\sigma(x+b_1-a_1-a_2-e)}{\sigma(x-a_1)\sigma(x-a_2)} \exp(-\eta e \int^x dz), \qquad \lambda \in ¢^*$$

is a meromorphic section of $e \in J_0$ such that

$$\text{(ii)} \qquad r = \frac{h(b)}{h(a)} = \frac{\sigma(a-a_1)\sigma(a-a_2)}{\sigma(b-a_1)\sigma(b-a_2)} \exp(m + \eta \int_{a_1+a_2}^{a+b} v) \int_b^a v \in ¢^*,$$

with η given by (46) and $\int_{a_1+a_2}^{a+b} v - e = \begin{Bmatrix} m \\ * \end{Bmatrix}_\tau \in ¢$, $m \in \mathbb{Z}$. Thus, there will be a solution $\mathcal{B} \not\ni a,b$ to (★) in the exceptional case if and only if r satisfies condition (ii), in which case there is a one parameter family of solutions provided by the zeroes of (i) as b_1 varies over C.

IV. Cyclic Unramified Coverings

One of the high points of classical theta-function theory was Schottky's discovery that his modular form on $\mathcal{H}_4$ describing the locus of period matrices of Riemann surfaces of genus 4 [28] could be expressed in terms of nullwerthe of genus 3 theta functions associated to the Prym varieties for double unramified coverings of genus 4 curves. Although Riemann [27], Wirtinger [34] and Schottky-Jung [29] all worked on different aspects of the same theory, their diverse approaches to the problem made it difficult for a unified theory to emerge. In this chapter, their various results are presented under a general discussion of theta relations on curves of special moduli admitting a fixed-point-free automorphism of order $p \geq 2$.

Relations between $\hat{\Theta}$ and Θ functions. Suppose that $\hat{C} \overset{\pi}{\to} C$ is a Riemann surface of genus $\hat{g} = pg-p+1$ given as an unramified cyclic covering of degree $p \geq 2$ of a compact Riemann surface C of genus $g \geq 1$. Let $\phi\colon \hat{C} \to \hat{C}$ be a conformal automorphism of $\hat{C}$ generating the group $G = \{\phi^n,\ n = 0,\dots,p-1\}$ of cover transformations of $\hat{C} \to C$, and write $D^{(n)} = \phi^n(D)$ for any divisor D on $\hat{C}$. The map $\pi^*\colon J(C) \to J(\hat{C})$, lifting divisors of degree 0, has as kernel a cyclic subgroup of order p of the group of points of order p in J(C), and the generator of this subgroup can be identified with the characteristic homomorphism of the discrete bundle $\hat{C} \to C$. Since any p^{th}-integer period characteristic is equivalent under $Sp(2g,\mathbb{Z})$ to one of the form $\begin{bmatrix} 0 & 0 & \cdots & 0 \\ \frac{k}{p} & 0 & \cdots & 0 \end{bmatrix}$ with $0 \leq k \leq p-1$, we can assume that a canonical homology basis $A_0,B_0,A_1,B_1,\dots,A_{g-1},B_{g-1}$ of $H_1(C,\mathbb{Z})$ is chosen so that the covering is defined, say, by the characteristic

$\varepsilon = \begin{bmatrix} 0 & 0 & \cdots & 0 \\ \frac{1}{p} & 0 & \cdots & 0 \end{bmatrix}$. With such a basis for $H_1(C,\mathbb{Z})$ there is a particularly simple choice of a basis for $H_1(\hat{C},\mathbb{Z})$ such that the corresponding lattice subgroup in $\mathbb{C}^{\hat{g}}$ defining $J(\hat{C})$ will give rise to a $\hat{\Theta}$-divisor invariant under the action of G on $J(\hat{C})$. Just pick a basis of $H_1(\hat{C},\mathbb{Z})$:

$\hat{A}_0, \hat{B}_0, \hat{A}_1, \hat{B}_1, \ldots, \hat{A}_{g-1}, \hat{B}_{g-1}, \ldots, \hat{A}_{\hat{g}-1}, \hat{B}_{\hat{g}-1}$

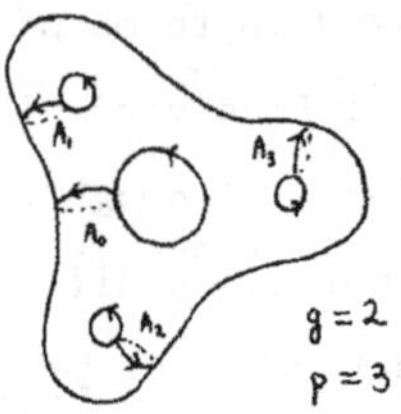

such that $\pi\hat{A}_0 = A_0$, $\pi\hat{B}_0 = pB_0$,

$$\bigcup_{n=0}^{p-1} \hat{A}_{k+n(g-1)} = \pi^{-1}(A_k), \qquad \bigcup_{n=0}^{p-1} \hat{B}_{k+n(g-1)} = \pi^{-1}(B_k)$$

and

$\hat{A}_{k+n(g-1)} = \phi^n(\hat{A}_k)$, $\hat{B}_{k+n(g-1)} = \phi^n(\hat{B}_k)$ for $1 \le k \le g-1$. The corresponding normalized differentials $u_0, \ldots, u_{\hat{g}-1}$ on $\hat{C}$ then satisfy

$$(58) \qquad (\phi^n)^* u_0 = u_0 \quad \text{and} \quad (\phi^n)^* u_{n(g-1)+k} = u_k, \qquad 1 \le k \le g-1, \quad 0 \le n \le p-1$$

and the Riemann matrix $\hat{\tau}$ for $\hat{C}$ has the symmetric form

$$(59) \qquad \hat{T} = \left(\begin{array}{c|c|c|c|c} \hat{\tau}_{00} & \hat{\tau}_0 & \hat{\tau}_0 & \cdots & \hat{\tau}_0 \\ \hline \hat{\tau}_0^t & M & M' & \cdots & M^{(p-1)} \\ \hline & & & & \\ \vdots & \vdots & \vdots & & \vdots \\ \hline \hat{\tau}_0^t & M' & M'' & \cdots & M \end{array}\right)$$

where $M^{(n)} = \left(\int_{\phi^n(\hat{B}_j)} u_i\right)$ and $\hat{\tau}_0 = \left(\int_{\hat{B}_0} u_j\right)$ for $1 \le i,j \le g-1$.

The normalized differentials on C are given by

$$(60) \qquad v_0 = u_0, \quad v_i = u_i + \phi^* u_i + \ldots + (\phi^{p-1})^* u_i \qquad 1 \le i \le g-1$$

with period matrix

$$\tau_{00} = \frac{1}{p}\hat{\tau}_{00} \qquad \tau_{0j} = \hat{\tau}_{0j} \qquad 1 \le j \le g-1$$

$$\tau_{ij} = \hat{\tau}_{ij} + \hat{\tau}_{i,j+g-1} + \ldots + \hat{\tau}_{i,j+(p-1)(g-1)} \qquad 1 \le i,j \le g-1.$$

From (60) and (30), the differentials of the second kind $\omega(x,y)$ and $\hat{\omega}(x,y)$ corresponding to the choice of homology bases for C and $\hat{C}$ satisfy

$$(61) \qquad \omega(x,y) = \hat{\omega}(x,y) + \hat{\omega}(x,y') + \ldots + \hat{\omega}(x,y^{(p-1)}) \qquad \forall\ x,y \in \hat{C}.$$

Finally, there is a decomposition

$$(62) \qquad H^0(\Omega^1_{\hat{C}}) \simeq \bigoplus_{n=0}^{p-1} H^0(\underline{n\varepsilon} \otimes \Omega^1_C)$$

where $\underline{n\varepsilon}$ is the sheaf of sections of the flat line bundle defined by the characteristic $\begin{bmatrix} 0 & \ldots & 0 \\ \frac{n}{p} & \ldots & 0 \end{bmatrix}$. Elements of $H^0(\underline{n\varepsilon} \otimes \Omega^1_C)$ are the Prym differentials on C with multipliers $n\varepsilon$; for each $n \neq 0$, there are $g-1$ such linearly independent differentials and a basis is given by

$$u_j + e^{\frac{2\pi i n}{p}}(\phi)^* u_j + \ldots + e^{\frac{2\pi i n}{p}(p-1)}(\phi^{p-1})^* u_j \qquad 1 \leq j \leq g-1.$$

The mapping $\pi^*\colon J(C) \to J(\hat{C})$ can be lifted to a map from $\mathbb{C}^g$ to $\mathbb{C}^{\hat{g}}$, which will again be denoted by π^*: with the above choice of homology on C and $\hat{C}$,

$$\pi^* z = \pi^*(z_0, z_1, \ldots, z_{g-1}) = (pz_0, z_1, \ldots, z_{g-1}, \ldots, z_1, \ldots, z_{g-1}) \in \mathbb{C}^{\hat{g}}$$

for $z \in \mathbb{C}^g$; in terms of period characteristics, (59)-(60) give:

$$(63) \qquad \pi^* \begin{Bmatrix} \alpha^0 & \alpha \\ \beta^0 & \beta \end{Bmatrix}_\tau = \begin{Bmatrix} \alpha^0 & \alpha & \ldots & \alpha \\ p\beta^0 & \beta & \ldots & \beta \end{Bmatrix}_{\hat{\tau}} \in \mathbb{C}^{\hat{g}}$$

for all $\alpha^0, \beta^0 \in \mathbb{R}$ and $\alpha, \beta \in \mathbb{R}^{g-1}$. Likewise the action of G on $J(\hat{C})$ gives rise to a group of automorphisms of $\mathbb{C}^{\hat{g}}$ by (58):

$$\phi^{p-n}(\hat{z}) = (\hat{z}_0, \hat{z}_{1+n(g-1)}, \ldots, \hat{z}_{g-1+n(g-1)}, \ldots, \hat{z}_{1+(n-1)(g-1)}, \ldots, \hat{z}_{g-1+(n-1)(g-1)}) \in \mathbb{C}^{\hat{g}}$$

for $\hat{z} \in \mathbb{C}^{\hat{g}}$ and $0 \leq n \leq p-1$; in characteristic notation:

$$(64) \qquad \phi^{p-n}\begin{Bmatrix} \alpha^0 & \alpha^1 & \dots & \alpha^p \\ \beta^0 & \beta^1 & \dots & \beta^p \end{Bmatrix}_{\hat{\tau}} = \begin{Bmatrix} \alpha^0 & \alpha^{n+1} & \dots & \alpha^n \\ \beta^0 & \beta^{n+1} & \dots & \beta^n \end{Bmatrix}_{\hat{\tau}} \in \mathbb{C}^{\hat{g}}$$

for $\alpha^0, \beta^0 \in \mathbb{R}$ and $\alpha^k, \beta^k \in \mathbb{R}^{g-1}$, $1 \leq k \leq p$. If we define the Riemann form for $\hat{\tau}$ by

$$(65) \qquad H(\hat{z},\hat{w}) = -\frac{1}{2\pi} \sum_{i,j=1}^{\hat{g}} (\mathrm{Re}\ \hat{\tau})^{-1}_{ij}\ \hat{z}_i \overline{\hat{w}_j}\ , \qquad \hat{z},\hat{w} \in \mathbb{C}^{\hat{g}}$$

then H is a positive definite form on $\mathbb{C}^{\hat{g}}$ and Im H has integral values on the lattice in $\mathbb{C}^{\hat{g}}$ defined by the columns of $(2\pi i I, \hat{\tau})$ [33]; from (59) and (64), H has the invariance property

$$H(\hat{z},\hat{w}) = H(\phi^n(\hat{z}),\phi^n(\hat{w})) \quad \forall\ \hat{z},\hat{w} \in \mathbb{C}^{\hat{g}}, \qquad 0 \leq n \leq p-1.$$

Similarly, the function $\hat{\Theta}$ constructed from $\hat{\tau}$ has the symmetries

$$(66) \qquad \hat{\Theta}(\hat{z}) = \hat{\Theta}(\phi^n(\hat{z})) \qquad \forall\ \hat{z} \in \mathbb{C}^{\hat{g}} \quad \text{and} \quad 0 \leq n \leq p-1.$$

This equation, together with Theorem 1.1 (10), implies that the Riemann divisor class $\hat{\Delta} \in J_{\hat{g}-1}(\hat{C})$ corresponding to $\hat{\Theta}$ is a fixed point under the action of G on $J_{\hat{g}-1}(\hat{C})$.

<u>Proposition 4.1.</u>* Let λ be the half-period in $J(\hat{C})$ corresponding to the bundle $\hat{L}_0 \otimes \pi^*(L_0^{-1})$ of order 2 on $\hat{C}$. Then for all $z \in \mathbb{C}^g$,

$$(67) \qquad \frac{\hat{\Theta}[\lambda](\pi^* z)}{\prod_{k=0}^{p-1} \Theta\begin{bmatrix} 0 & \dots & 0 \\ \frac{k}{p} & \dots & 0 \end{bmatrix}(z)} = c, \text{ a non-zero constant,}$$

so that $\pi^*(\Theta) = (\hat{\Theta}[\lambda]) \cap \pi^* J(C)$. In terms of the above choice of homology basis for $\hat{C}$, $\lambda = \pi^*(\frac{p-1}{p}\pi i, 0, \dots, 0) \in J(\hat{C})$.

---

* This theorem was observed by Mumford and also proved by Accola [1,I, p. 33].

Proof. Let $f = \zeta - \Delta$ be a (generic) non-singular point of (Θ), with ζ a positive divisor of degree $g-1$ on C; then

$$\pi^* f = \pi^*\zeta - \pi^*\Delta = \hat{\zeta} - \hat{\Delta} + \lambda \in J(\hat{C})$$

where $\hat{\zeta} = \pi^*\zeta$ is of degree $p(g-1) = \hat{g}-1$ and $\lambda = \hat{\Delta} - \pi^*\Delta \in J_0(\hat{C})$ is a half-period since by (9):

$$2\pi^*\Delta = \pi^*(2\Delta) = \pi^* K_C = K_{\hat{C}} = 2\hat{\Delta}.$$

Similarly if $f + (\frac{2\pi ik}{p},0,\ldots,0) \in (\Theta)$, $\pi^* f - \lambda \in (\hat{\Theta})$; therefore $\hat{\Theta}[\lambda](\pi^* z) / \prod_{k=0}^{p-1} \Theta\begin{bmatrix} 0 & \ldots & 0 \\ \frac{k}{p} & \ldots & 0 \end{bmatrix}(z)$ is a holomorphic section of a flat line bundle on J(C) since it is infinite at most on a $g-3$ dimensional set in J(C) corresponding to the singular set of $\Theta\begin{bmatrix} 0 & \ldots & 0 \\ \frac{k}{p} & \ldots & 0 \end{bmatrix}$ for $k = 1,\ldots,p-1$. Since λ is a half-period, it has to remain constant for a family $\wp$ of curves $\hat{C}_t$ covering a family of curves C_t over the unit disc which are being pinched along the cycle $A_0 B_0 A_0^{-1} B_0^{-1}$ as $t \to 0$. By (59) and Cor. 3.2, the matrix $\hat{\tau}(t)$ for C_t degenerates to a sum of $p+1$ blocks along the diagonal; so the condition that $\hat{\Theta}_{\hat{\tau}(t)}[\lambda](\pi^* z) / \prod_{k=0}^{p-1} \Theta\begin{bmatrix} 0 & \ldots & 0 \\ \frac{k}{p} & \ldots & 0 \end{bmatrix}_{\tau(t)}(z)$ has no poles implies, as $t \to 0$, that $[\lambda] = \begin{bmatrix} \alpha_0 & 0 & \ldots & 0 \\ \beta_0 & 0 & \ldots & 0 \end{bmatrix}$ where $2\alpha_0, 2\beta_0 \in \mathbf{Z}$ are such that $\Theta_{p\tau_{00}}\begin{bmatrix} \alpha_0 \\ \beta_0 \end{bmatrix}(pz_0) / \prod_{0}^{p-1} \Theta_{\tau_{00}}\begin{bmatrix} 0 \\ \frac{k}{p} \end{bmatrix}(z_0)$ is holomorphic $\forall\ z_0 \in \mathbb{C}$. Now a formula from elliptic functions [19, p. 402, #122] gives $\begin{bmatrix} \alpha_0 \\ \beta_0 \end{bmatrix} = \begin{bmatrix} 0 \\ \frac{p-1}{2} \end{bmatrix}$ for which the above expression is a constant. So $[\lambda] = \begin{bmatrix} 0 & 0 & \ldots & 0 \\ \frac{p-1}{2} & 0 & \ldots & 0 \end{bmatrix}_{\hat{\tau}}$ and $\hat{\Theta}[\lambda](\pi^* z) / \prod_{0}^{p-1} \Theta\begin{bmatrix} 0 & \ldots & 0 \\ \frac{k}{p} & \ldots & 0 \end{bmatrix}(z)$ is a constant which is non-zero since the multiplicity of $\hat{\Theta}[\lambda]$ at $\pi^* z$ is the sum of the multiplicities of $\Theta\begin{bmatrix} 0 & \dots & 0 \\ \frac{k}{p} & \dots & 0 \end{bmatrix}$ at z by virtue of (62):

$$i(\hat{\zeta}) = \dim H^1(\hat{C},\hat{\zeta}) = \sum_{k=0}^{p-1} \dim H^1(C\,,\,\zeta \otimes k\varepsilon) = \sum_{k=0}^{p-1} i_k(\zeta)$$

where $i_k(\zeta)$, the multiplicity of $\Theta\begin{bmatrix} 0 & \dots & 0 \\ \frac{k}{p} & \dots & 0 \end{bmatrix}$ at $z = \zeta-\Delta$, is the number of linearly independent differentials with multipliers $k\varepsilon$ which vanish at ζ.

We will retain the notation $E(x,y)$ for the prime-form on $C \times C$ lifted to $\hat{C} \times \hat{C}$.

Corollary 4.2. For fixed $x \in \hat{C}$, $\dfrac{E(x,y)}{\hat{E}(x,y)}$ is a multiplicative holomorphic function on $\hat{C}$ with $p-1$ zeroes at $y = x', x'', \dots, x^{(p-1)}$ and is a section of the bundle $\pi^*\delta^*(\Theta) \otimes \hat{\delta}^*(\hat{\Theta})^{-1} \otimes \lambda$ with, as on p. 16, $\delta(y) = \pi_*(y-x) \in J(C)$ and $\hat{\delta}(y) = y-x \in J(\hat{C})$.

Corollary 4.3. For $x \in \hat{C}$, $f \in (\Theta)$ and c as in Prop. 4.1,

$$\sum_1^{\hat{g}} \frac{\partial\hat{\Theta}[\lambda]}{\partial\hat{z}_i}(\pi^* f)u_i(x) = \frac{c}{p}\prod_{k=1}^{p-1}\Theta\begin{bmatrix} 0 & \dots & 0 \\ \frac{k}{p} & \dots & 0 \end{bmatrix}(f)\sum_1^{g}\frac{\partial\Theta}{\partial z_i}(f)v_i(x).$$

Proof. Use (67) and the fact that, from the symmetries of $\hat{\Theta}$ and λ:

$$\frac{\partial\hat{\Theta}[\lambda]}{\partial y}\left(\pi^*\left(\int_x^y v + f\right)\right)dy\Big|_{y=x} = p\sum_0^{g-1}\frac{\partial\hat{\Theta}[\lambda]}{\partial\hat{z}_i}(\pi^* f)v_i(x) = p\sum_0^{\hat{g}-1}\frac{\partial\hat{\Theta}[\lambda]}{\partial\hat{z}_i}(\pi^* f)u_i(x).$$

Corollary 4.4. Let f be a non-singular point on (Θ) with $f+k\varepsilon \notin (\Theta)$ for $k = 1,\dots,p-1$. Then $\forall\ x,y \in \hat{C}$,

$$\frac{\hat{\Theta}[\lambda]\left(\int_x^y u - \pi^* f\right)}{\Theta\left(\int_x^y v - f\right)\prod_1^{p-1}\Theta\begin{bmatrix} 0 & \dots & 0 \\ \frac{k}{p} & \dots & 0 \end{bmatrix}(f)} = \frac{c}{p}\,\frac{\hat{E}(x,y)}{E(x,y)}.$$

Proof. Write $f = \zeta-\Delta$ with ζ a positive divisor of degree $g-1$;

then for fixed x, $\mathrm{div}_{\hat{C}}\,\Theta(\int_x^y v - f) = \pi^*\zeta + \pi^* x$ and $\mathrm{div}_{\hat{C}}\hat{\Theta}[\lambda](\int_x^y u - \pi^* f) =$ $x + \hat{\zeta}$ where $\hat{\zeta} = \pi^* f + \hat{\Delta} + \lambda = \pi^*(f + \Delta) = \pi^*\zeta$. By Corollary 4.2,

$$\frac{E(x,y)}{\hat{E}(x,y)} \frac{\hat{\Theta}[\lambda](\int_x^y u - \pi^* f)}{\Theta(\int_x^y v - f)}$$

is therefore a well-defined function on C with no zeroes or poles; by letting $y \to x$ and using Cor. 4.3, this constant is found to be $\frac{c}{p} \prod_1^{p-1} \Theta\begin{bmatrix} 0 & \dots & 0 \\ \frac{k}{p} & \dots & 0 \end{bmatrix}(f)$.

Proposition 4.5.* Let $e \in \mathbb{C}^g$ with $e + k\varepsilon \notin (\Theta)$ for $k = 0,1,\dots,p-1$. Then $\forall\, x,y \in \hat{C}$,

$$p\, \frac{E(x,y)}{\hat{E}(x,y)} \frac{\hat{\Theta}[\lambda](\int_x^y u - \pi^* e)}{\hat{\Theta}[\lambda](\pi^* e)} = \sum_{k=0}^{p-1} \frac{\Theta\begin{bmatrix} 0 & \dots & 0 \\ \frac{k}{p} & \dots & 0 \end{bmatrix}(\int_x^y v - e)}{\Theta\begin{bmatrix} 0 & \dots & 0 \\ \frac{k}{p} & \dots & 0 \end{bmatrix}(e)}.$$

Proof. Since $\Theta(e + k\varepsilon) \neq 0$ for $0 \leq k \leq p-1$, $\Theta\begin{bmatrix} 0 & \dots & 0 \\ \frac{k}{p} & \dots & 0 \end{bmatrix}(\int_x^y v - e) \Big/ \Theta(\int_x^y v - e)$ are p linearly independent functions of $y \in \hat{C}$ (with x fixed), forming a basis for the functions on $\hat{C}$ as a vector space over the functions on C. But by Cor. 4.2,

$$\frac{E(x,y)}{\hat{E}(x,y)} \frac{\hat{\Theta}[\lambda](\int_x^y u - \pi^* e)}{\Theta(\int_x^y v - e)}$$

is a function on $\hat{C}$; so we can write

$$\frac{E(x,y)}{\hat{E}(x,y)} \hat{\Theta}[\lambda](\int_x^y u - \pi^* e) = \sum_{k=0}^{p-1} \psi_k(x,y)\Theta\begin{bmatrix} 0 & \dots & 0 \\ \frac{k}{p} & \dots & 0 \end{bmatrix}(\int_x^y v - e)$$

where the $\psi_k(x,y)$ are meromorphic functions on $C \times C$. Replacing y by $y', y'', \dots, y^{(p-1)}$ we see that each $\psi_k(x,y)$ can have poles only at the

* This appears in [29] for $p = 2$ - see Prop. 4.19.

zeroes of $\Theta\begin{bmatrix}0 & \cdots & 0\\ \frac{k}{p} & \cdots & 0\end{bmatrix}(\int_x^y v - e)$ so that $\psi_k(x,y)$ is constant in y since the index of speciality of $\mathrm{div}_C\,\Theta\begin{bmatrix}0 & \cdots & 0\\ \frac{k}{p} & \cdots & 0\end{bmatrix}(\int_x^y v - e)$ is zero. By replacing y with $x, x', \ldots, x^{(p-1)}$, we conclude that $\psi_k(x,y) = \frac{1}{p}\hat{\theta}[\lambda](\pi^* e)/\Theta\begin{bmatrix}0 & \cdots & 0\\ \frac{k}{p} & \cdots & 0\end{bmatrix}(e)$ for all $x,y \in C$.

<u>The Prym Variety</u>. The Prym variety P for the covering $\hat{C} \overset{\pi}{\to} C$ is the subvariety of $J(\hat{C})$ given by

$$P = \{n_0\mathcal{A} + n_1\mathcal{A}' + \ldots + n_{p-1}\mathcal{A}^{(p-1)} \mid \mathcal{A} \in J_0(\hat{C}),\ n_i \in \mathbb{Z},\ \sum_0^{p-1} n_i = 0\}.$$

Equivalently,

$$(68)\quad P = \{\mathcal{A} - \mathcal{A}' \mid \mathcal{A} \in J_0(\hat{C})\} = \{p\mathcal{B} \in J_0(\hat{C}) \mid \mathcal{B} + \mathcal{B}' + \ldots + \mathcal{B}^{(p-1)} = 0\}.$$

<u>Proposition 4.6</u>. The group P is isomorphic to $J(\hat{C})/\pi^*J(C)$ and there is an isogeny* $i: J(C) \times P \to J(\hat{C})$ of degree p^{2g-1}. The universal cover $\tilde{P}$ of P is the orthocomplement of the universal cover $\widetilde{\pi^*J}$ of π^*J in $\mathbb{C}^{\hat{g}}$ under the G-invariant Riemann form H given by (65).

<u>Proof</u>. The projection $\sigma: J(\hat{C}) \to P$, defined by $\sigma(\mathcal{A}) = \mathcal{A} - \mathcal{A}'$ for $\mathcal{A} \in J_0(\hat{C})$, has kernel $\pi^*J(C)$ by (63)-(64); so $J(\hat{C})/\pi^*J(C) \simeq P$ under σ and P has dimension $\hat{g}-g = (p-1)(g-1)$. From (64) and (68), P is the subgroup of $J(\hat{C})$ given by all points with characteristics

$$(69)\quad \begin{Bmatrix}0 & \alpha^1 & \cdots & \alpha^p\\ 0 & \beta^1 & \cdots & \beta^p\end{Bmatrix}_{\hat{\tau}} \in J(\hat{C}), \qquad \sum_1^p \alpha^k \text{ and } \sum_1^p \beta^k \in \mathbb{Z}^{g-1},$$

so that by (63), $\pi^*J \cap P$ is isomorphic to the group of $\hat{\tau}$-characteristics of the form $\pi^*\begin{bmatrix}0 & \delta_1 & \cdots & \delta_{g-1}\\ 0 & \varepsilon_1 & \cdots & \varepsilon_{g-1}\end{bmatrix}_\tau$, $p\delta_i$ and $p\varepsilon_i \in \mathbb{Z}$, a group of

---

* "isogeny" here means group epimorphism with finite kernel.

order $p^{2(g-1)}$; thus if the isogeny $i: J(C) \times P \to J(\hat{C})$ is defined by $i(A,B) = \pi^*A + B$ for $A \in J$ and $B \in P$, then kernel $i \cong$ kernel $\pi^* \times (\pi^*J \cap P)$ is a group of order $p^{2(g-1)} \dim \ker \pi^* = p^{2g-1}$, the degree of the isogeny. For $Z \in \widetilde{\pi^*J}$ and $W \in \tilde{P}$, the symmetry of H implies

$$0 = H(Z, \sum_0^{p-1} \phi^n W) = H(Z,W) + H(\phi Z, \phi W) + \ldots + H(\phi^{p-1}Z, \phi^{p-1}W) = pH(Z,W)$$

so that $\tilde{P}$ and $\widetilde{\pi^*J}$ are orthogonal under H. Conversely, if $H(Z,W) = 0$ for some $W \in \mathbb{C}^{\hat{g}}$ and all $Z \in \widetilde{\pi^*J}$, then $H(Z, \sum_0^{p-1} \phi^n W) = 0$; taking $Z = \sum_0^{p-1} \phi^n W$, the positivity of H implies $\sum_0^{p-1} \phi^n W = 0$ and $W \in \frac{1}{p}\tilde{P} = \tilde{P}$.

<u>Proposition 4.7.</u> For $n = 0,1,\ldots,p-1$, let $\delta_n = n\pi^*\left(\frac{\tau_{00}}{p}, \frac{\tau_{01}}{p}, \ldots, \frac{\tau_{0\,g-1}}{p}\right)$ be a point of $J(\hat{C})$ such that $\phi^n(x) - x \in P + \delta_n$ for any $x \in \hat{C}$. Then if $\pi_*(\mathcal{A})$ is the projection to C of a divisor $\mathcal{A}$ on $\hat{C}$,

$$\{\mathcal{A} \in J(\hat{C}) \mid \pi_*(\mathcal{A}) = 0 \text{ in } J(C)\} = \coprod_{n=0}^{p-1} (P + \delta_n)$$

and

$$\coprod_{n=0}^{p-1} \operatorname{div} \hat{\theta}[\lambda] \cap (P+\delta_n) = \{\mathcal{B} - \pi^*\Delta \mid \mathcal{B} \text{ is positive of degree } \hat{g}-1 \text{ and } \pi_*(\mathcal{B}) = p\Delta\}$$

<u>Proof.</u> We first show that if $\varepsilon_m = \pi^*\left(\frac{2\pi i m}{p^2}, 0, \ldots, 0\right)$, then

$$\{\mathcal{A} \in J(\hat{C}) \mid \sum_{k=0}^{p-1} \mathcal{A}^{(k)} = 0 \text{ in } J(\hat{C})\} = \coprod_{m,n=0}^{p-1} (P + \delta_n + \varepsilon_m).$$

To see this, write $\mathcal{A} = A + B$ with $A \in \pi^*J$ and $B \in P$; then $\sum_{k=0}^{p-1} \mathcal{A}^{(k)} = 0$ implies that $\sum_0^{p-1} A^{(k)} = pA = 0$ so that A is contained in the group of points of order p in π^*J. From the description of

$\pi^*J \cap P$ given in the proof of Prop. 4.6, $A - \delta_n - \varepsilon_m \in \pi^*J \cap P$ for some integers m and n so that $\mathcal{A} \in P + \delta_n + \varepsilon_m$, as asserted. Now if $\pi_*(\mathcal{A}) = 0$ for $\mathcal{A} \in J(\hat{C})$, $\sum_{k=0}^{p-1} \mathcal{A}^{(k)} = 0$ in $J(\hat{C})$ so that $\mathcal{A} \in P + \delta_n + \varepsilon_m$ for some integers m,n; but if $0 < m < p$, $\pi_*(\mathcal{A}) = (\frac{2\pi i m}{p}, 0, 0, \ldots, 0) \neq 0$ in J(C), so actually we must have $\mathcal{A} \in P + \delta_n$. Finally, $f \in \operatorname{div} \Theta[\lambda] \cap (P + \delta_n)$ if and only if $f = \mathcal{B} - \pi^*\Delta$ with $\mathcal{B}$ positive of degree $\hat{g}-1$ by (11) and

$$0 = \pi_* f = \pi_* \mathcal{B} - \pi_*(\pi^*\Delta) = \pi_*\mathcal{B} - p\Delta.$$

Let $\sigma_1 : J(\hat{C}) \to J(C)$ and $\sigma_2 : J(\hat{C}) \to P$ be the projections defined by $\sigma_1(\mathcal{A}) = \pi_*(\mathcal{A}) = \pi(\mathcal{A})$ and $\sigma_2(\mathcal{A}) = (p-1)\mathcal{A} - \mathcal{A}' - \ldots - \mathcal{A}^{(p-1)}$ for $\mathcal{A} \in J(\hat{C})$. Then σ_1 and σ_2 satisfy $\pi^*\sigma_1 + \sigma_2 = p\,\mathrm{Id}_{J(\hat{C})}$, a relation which can be lifted to $\mathbb{C}^{\hat{g}}$ in the following manner: for $\hat{z} \in \mathbb{C}^{\hat{g}}$, let

$$z = \sigma_1(\hat{z}) = (\hat{z}_0, z_1, \ldots, z_{g-1}) \in \mathbb{C}^g \qquad \text{with } z_i = \sum_{n=0}^{p-1} \hat{z}_{i+n(g-1)},$$

and set

$$\hat{s} = \sigma_2(\hat{z}) = (0, -s^0, -s^1, \ldots, -s^{p-1}) \in \mathbb{C} \oplus (\mathbb{C}^{g-1})^p = \mathbb{C}^{\hat{g}}$$

where, by (64)

$$s^k = (-p\hat{z}_{i+k(g-1)} + \sum_0^{p-1} \hat{z}_{i+n(g-1)})_{1 \le i \le g-1} \in \mathbb{C}^{g-1}, \qquad \sum_0^{p-1} s^k = 0 \in \mathbb{C}^{g-1}.$$

If $\tilde{P} \subset \mathbb{C}^{\hat{g}}$ is the universal cover of $P \subset J(\hat{C})$, there is an isomorphism $\phi : \mathbb{C}^{\hat{g}-g} = (\mathbb{C}^{g-1})^{p-1} \cong \tilde{P}$ defined by

$$(70) \qquad s = (s^1, \ldots, s^{p-1}) \in (\mathbb{C}^{g-1})^{p-1} \longrightarrow \hat{s} = \phi(s) = (0, \sum_1^{p-1} s^k, -s^1, \ldots, -s^{p-1}) \in \mathbb{C}^{\hat{g}}$$

Then $\forall\, \hat{z} \in \mathbb{C}^{\hat{g}}$,

$$(71) \qquad p\hat{z} = \pi^*\sigma_1(\hat{z}) + \sigma_2(\hat{z}) = \pi^*z + \hat{s} = \pi^*z + \phi(s) \in \mathbb{C}^{\hat{g}}.$$

Proposition 4.8. There is a principally polarized Abelian variety P_0, an isogeny $i: P_0 \to P$ of degree $p^{(p-2)(g-1)}$ and a projection $\tilde{\sigma}_2: J(\hat{C}) \to P_0$ such that $i\circ\tilde{\sigma}_2 = \sigma_2$ and

$$(72) \qquad p(\hat{\Theta}) \sim \sigma_1^{-1}(\Theta) + \tilde{\sigma}_2^{-1}(\eta) \qquad \text{on } J(\hat{C})\ ^*$$

where (η) is the theta-divisor on P_0.

Proof. We will make an explicit construction of P_0 as a principally polarized Abelian variety $\mathbb{C}^{\hat{g}-g}/(2\pi i I,\Pi)$ with Π a symmetric $(\hat{g}-g)\times(\hat{g}-g)$ matrix with negative definite real part; the isogeny i will then be induced from the isomorphism ϕ given by (70), and the assertion (72) that

$$(72)' \qquad \hat{\theta}^p_{\hat{\tau}}(\hat{z})/\theta_\tau(\sigma_1(\hat{z}))\eta_\pi(\phi^{-1}\sigma_2(\hat{z})) \qquad \hat{z} \in \mathbb{C}^{\hat{g}}$$

is a meromorphic function on $J(\hat{C})$ will then be a consequence of the identities connecting the $\hat{\Theta}$, Θ and η-functions, the latter constructed from Π. So let $\psi: \mathbb{C}^{\hat{g}-g} \xrightarrow{\sim} \tilde{P}$ be the isomorphism defined by sending $s = (s^1,\dots,s^{p-1}) \in \mathbb{C}^{\hat{g}-g}$ into

$$\psi(s) = (0,s^1+..+s^{p-1},(1-p)s^1+s^2+..+s^{p-1},\dots,s^1+s^2+..+(1-p)s^{p-1} \in \mathbb{C}^{\hat{g}}\ ;$$

for $s^k \in \mathbb{C}^{g-1}$; then ϕ and ψ satisfy

$$(73) \qquad p\,X\cdot Y = \phi(X)\cdot\psi(Y) \qquad X,Y \in \mathbb{C}^{\hat{g}-g}$$

and $\phi\psi^t(\hat{s}) = p\hat{s}$ for any $\hat{s} \in \tilde{P}$. The quadratic form $\hat{Q}$ on $\mathbb{C}^{\hat{g}}$ defined by the matrix $\hat{\tau}$ can, restricted to $\tilde{P} \subset \mathbb{C}^{\hat{g}}$, be considered a quadratic form $\tilde{Q}$ on $\mathbb{C}^{\hat{g}-g}$ by pulling back by ψ; if Ψ is the matrix defining ψ, the matrix representing $\tilde{Q}$ on $\mathbb{C}^{\hat{g}-g}$ is given by $p\Pi = \Psi^t\hat{\tau}\Psi$ with $\mathrm{Re}\,\Pi < 0$ since $\mathrm{Re}\,\hat{\tau} < 0$ on $\tilde{P}$. In terms of the matrices $M^{(n)}$ in (59),

---

* $\sim$ is linear equivalence of divisors on J.

$$\Pi = \begin{pmatrix} N & N' & N'' & \cdots & N^{p-2} \\ N^{p-1} & N & N' & \cdots & N^{p-3} \\ \vdots & \vdots & \vdots & & \vdots \\ & & & & N' \\ N'' & N''' & N^{iv} & \cdots & N \end{pmatrix}, \quad N^{(k)} = pM^k - \sum_{n=0}^{p-1} M^{(n)} .$$

If Γ is the lattice generated in $\mathbb{C}^{\hat{g}}$ by the columns of $\hat{\tau}$, the lattice generated by the columns of Π in $\mathbb{C}^{\hat{g}-g}$ is mapped by ϕ onto the sublattice $L_0 = \{(1-p)\gamma + \gamma' + \gamma^{(p-1)} \mid \gamma \in \Gamma\}$ of the lattice $L = \Gamma \cap \tilde{P} = \{\sum_0^{p-1} n_i \gamma^{(i)} \mid \gamma \in \Gamma,\ n_i \in \mathbf{Z},\ \sum n_i = 0\}$. On the other hand, the lattice generated by the columns of $(2\pi i I)$ in $\mathbb{C}^{\hat{g}-g}$ is mapped by ϕ onto the lattice Λ given by the intersection of $\tilde{P}$ with the lattice generated by the columns of $(2\pi i I)$ in $\mathbb{C}^{\hat{g}}$. Thus, under ϕ, $P_0 = \mathbb{C}^{\hat{g}-g}/(2\pi i I, \Pi)$ is isomorphic to $\tilde{P}/L_0+\Lambda$, and ϕ composed with the identity mapping on $\mathbb{C}^{\hat{g}}$ gives an isogeny $i: P_0 \to P = \tilde{P}/L+\Lambda$ of degree $p^{(p-2)(g-1)}$, the order of the group L/L_0. Furthermore, since the map σ_2 lifted to $\mathbb{C}^{\hat{g}}$ sends the lattice defining $J(\hat{C})$ in $\mathbb{C}^{\hat{g}}$ onto the sublattice $L_0+\Lambda$, $\sigma_2: J(\hat{C}) \to P$ can be lifted to a map $\tilde{\sigma}_2: J(\hat{C}) \to P_0$ such that $i \circ \tilde{\sigma}_2 = \sigma_2$.

If $m \in \mathbf{Z}^{\hat{g}}$, the orthogonality of $\pi^* J$ and P under the quadratic form $\hat{Q}$ gives

$$\hat{Q}(m) = \sum_{i,j=1}^{\hat{g}} \hat{\tau}_{ij} m_i m_j = pQ(n + \frac{\delta}{p}) + \tilde{Q}(\tilde{n} + \frac{\tilde{\delta}}{p})$$

where $Q(z) = \sum_{i,j=1}^{g} \tau_{ij} z_i z_j$ and $n \in \mathbf{Z}^g$, $\delta \in (\mathbf{Z}/p\mathbf{Z})^g$, $\tilde{n} \in \mathbf{Z}^{\hat{g}-g}$, $\tilde{\delta} \in (\mathbf{Z}/p\mathbf{Z})^{\hat{g}-g}$ are given by

$$(74) \quad \begin{cases} \frac{1}{p}\pi^* \sigma_1(m) = \pi^*(n + \frac{\delta}{p}), \qquad \frac{1}{p}\sigma_2(m) = \psi(\tilde{n} + \frac{\tilde{\delta}}{p}) \\ \pi^* \delta + \psi \tilde{\delta} \in (p\mathbf{Z})^{\hat{g}} \end{cases}$$

from (71). There is a $1-1$ correspondence between $m \in \mathbf{Z}^{\hat{g}}$ and such n, δ, $\tilde{n}$, $\tilde{\delta}$ satisfying (74), and hence

$$\hat{\theta}_{\hat{T}}(\hat{z}) = \sum_{m\in\mathbb{Z}^{\hat{g}}} \exp\left\{\tfrac{1}{2}\hat{Q}(m) + m\cdot\hat{z}\right\} = \sum_{m\in\mathbb{Z}^{\hat{g}}} \exp\left\{\tfrac{1}{2}\hat{Q}(m) + \tfrac{1}{p}\pi^*\sigma_1(m)\cdot\tfrac{1}{p}\pi^*\sigma_1(\hat{z}) + \tfrac{1}{p}\sigma_2(m)\cdot\tfrac{1}{p}\sigma_2(\hat{z})\right\}$$

$$= \sum_{\substack{n\in\mathbb{Z}^g,\ \tilde{n}\in\mathbb{Z}^{\hat{g}-g}\\ \delta\in(\mathbb{Z}/p\mathbb{Z})^g,\ \tilde{\delta}\in(\mathbb{Z}/p\mathbb{Z})^{\hat{g}-g}\\ \pi^*\delta+\psi\tilde{\delta}\in(p\mathbb{Z})^{\hat{g}}}} \exp\left\{\tfrac{1}{2}pQ(n+\tfrac{\delta}{p}) + (n+\tfrac{\delta}{p})\cdot z\right\} \exp\left\{\tfrac{1}{2}\tilde{Q}(\tilde{n}+\tfrac{\tilde{\delta}}{p}) + (\tilde{n}+\tfrac{\tilde{\delta}}{p})\cdot s\right\}$$

by (71), (73) and (74). So if η is the theta function on P_0 defined by Π,

$$\hat{\theta}_{\hat{T}}(\hat{z}) = \sum_{\tilde{\delta}\in(\mathbb{Z}/p\mathbb{Z})^{\hat{g}-g}} \theta_{pT}\begin{bmatrix}\frac{\delta}{p}\\ 0\end{bmatrix}(z)\,\eta_{p\Pi}\begin{bmatrix}\frac{\tilde{\delta}}{p}\\ 0\end{bmatrix}(s)$$

where $\delta \in (\mathbb{Z}/p\mathbb{Z})^g$ and $\pi^*\delta + \psi\tilde{\delta} \in (p\mathbb{Z})^{\hat{g}}$; more generally, by (71),

$$\text{(75)}\qquad \hat{\theta}_{\hat{T}}\begin{bmatrix}\mu\\ \nu\end{bmatrix}(\hat{z}) = \sum_{\substack{\tilde{\delta}\in(\mathbb{Z}/p\mathbb{Z})^{\hat{g}-g}\\ \pi^*\delta+\psi\tilde{\delta}\in(p\mathbb{Z})^{\hat{g}}}} \theta_{pT}\begin{bmatrix}\frac{\pi_*\mu+\delta}{p}\\ \pi_*\nu\end{bmatrix}(z)\,\eta_{p\Pi}\begin{bmatrix}\frac{\psi^{-1}\sigma_2(\mu)+\tilde{\delta}}{p}\\ \phi^{-1}\sigma_2(\nu)\end{bmatrix}(s)$$

for any $\mu,\nu \in \mathbb{R}^{\hat{g}}$. Replacing ν by $\nu+\varepsilon$ with $p\varepsilon \in (\mathbb{Z}/p\mathbb{Z})^{\hat{g}}$ and $\pi_*\varepsilon = 0$, this identity can be inverted to give:

$$\text{(76)}\qquad p^{(p-1)(g-1)}\,\theta_T\begin{bmatrix}\frac{\pi_*\mu}{p}\\ \pi_*\nu\end{bmatrix}(z)\,\eta_\Pi\begin{bmatrix}\frac{\psi^{-1}\sigma_2(\mu)}{p}\\ \phi^{-1}\sigma_2(\nu)\end{bmatrix}(s) = \sum_{\substack{p\varepsilon\in(\mathbb{Z}/p\mathbb{Z})^{\hat{g}}\\ \pi_*\varepsilon=0}} e^{-\frac{2\pi i}{p}\varepsilon\cdot\sigma_2(\mu)}\,\hat{\theta}_{\frac{\hat{T}}{p}}\begin{bmatrix}\mu\\ \nu+\varepsilon\end{bmatrix}(\hat{z})$$

for any $\mu,\nu \in \mathbb{R}^{\hat{g}}$. The exact statement (72)' follows by taking $\mu = \nu = 0$ in (76) and observing that $\hat{\theta}_{\frac{\hat{T}}{p}}\begin{bmatrix}0\\ \varepsilon\end{bmatrix}$ is a pth order theta-function for $J(\hat{C})$ with characteristics $\begin{bmatrix}0\\ 0\end{bmatrix}$ - see p. 2.

<u>Corollary 4.9</u>. If $e \in P$ and $a \in \hat{C}$, either $\eta(\tilde{\sigma}_2(x-a-e)) \equiv 0$ on $\hat{C}$ or $\mathrm{div}_{\hat{C}}\,\eta(\tilde{\sigma}_2(x-a-e)) = \zeta$ is of degree $(p-1)(\hat{g}-1) = p(p-1)(g-1)$ and satisfies

$$\text{(77)}\qquad pe = \zeta + \pi^*(\Delta + \pi_*a) - p(a+\hat{\Delta}) \in J_0(\hat{C}).$$

Furthermore $\pi_*(\zeta)$ is the divisor of a differential on C of order $\frac{p(p-1)}{2}$.

Proof. Setting $\mathrm{div}_C\,\theta(\sigma_1(x-a)) = \mathcal{B}$ and $\mathrm{div}_{\hat{C}}\,\hat{\theta}(x-a-e) = \mathcal{A}$, Prop. 4.8 gives $p\mathcal{A} = \pi^*\mathcal{B} + \zeta$; but by Riemann's Theorem, $0 = \pi^*(\mathcal{B} - \pi_*a - \Delta)$ and $e = \mathcal{A} - a - \hat{\Delta}$ so that

$$pe + pa + p\hat{\Delta} = p\mathcal{A} = \pi^*(\pi_*a + \Delta) + \zeta$$

which gives (77). Since $\pi_*\hat{\Delta} = \pi_*(\lambda + \pi^*\Delta) = p\Delta + \pi_*\lambda$ and $p\pi_*\lambda = 0$ by Prop. 4.1,

$$0 = \pi_*(pe) = \pi_*(\zeta) + p\pi_*(a) + p\Delta - p\pi_*(a) - p^2\Delta = \pi_*(\zeta) - (p^2-p)\Delta$$

and so $\pi_*(\zeta) = \frac{p(p-1)}{2} K_C$ by (9).

Double Unramified Coverings. When $\hat{C} \to C$ is an unramified double covering, the Prym variety $P_0 \overset{\phi}{\to} P$ is principally polarized with period matrix $\Pi = (\int_{B_j} w_i)$ for $1 \le i,j \le g-1$; here $w_1,\ldots,w_{g-1} \in H^0(K_C \otimes \varepsilon)$, with ε the half period $\begin{Bmatrix} 0 & 0 & \ldots & 0 \\ \frac{1}{2} & 0 & \ldots & 0 \end{Bmatrix}_\tau$, are $g-1$ linearly independent holomorphic Prym differentials on C normalized so that $(\frac{1}{2\pi i}\int_{A_j} w_i)$ is the identity $(g-1) \times (g-1)$ matrix. The simple symmetries involved make it easy to give very explicit relations between the $\hat{\theta}$, θ and η functions, basically due to the fact that the η-divisor on $\hat{C}$, of degree $2g-2$ by Corollary 4.9, is small enough to be effectively compared with the $\hat{\theta}[\lambda]$ divisor on $\hat{C}$ - see Prop. 4.14. To begin with, there is the following useful fact, due to Wirtinger [34, p. 90] and Mumford:

Proposition 4.10. For any divisor $\mathcal{B}$ on $\hat{C}$ with $\pi_*(\mathcal{B}) = K_C$, $i(\mathcal{B})$ is even if and only if $i(\mathcal{B} + x' - x)$ is odd for all $x \in \hat{C}$. Thus, by Prop. 4.7, $\mathrm{mult}_{\hat{z}}\,\hat{\theta}[\lambda]$ is even for $\hat{z} \in P + \delta_0 = P$ and odd for $\hat{z} \in P + \delta_1$.

Proof. We'll show that $i(\mathcal{A} + x) = k \in \mathbb{Z}$ if and only if $i(\mathcal{A} + x') = k \pm 1$ provided $\mathcal{A}$ satisfies $\mathcal{A} + \mathcal{A}' + x + x' = K_{\hat{C}}$. Now this last condition gives

$$1+\dim H^1(\mathcal{A}+x+x') = 1+\dim H^0(\mathcal{A}') = 1+\dim H^0(\mathcal{A}) = \dim H^1(\mathcal{A})$$

by the Riemann-Roch theorem. Since obviously $\dim H^1(\mathcal{A}) = \dim H^1(x+\mathcal{A})$ implies $\dim H^1(\mathcal{A}+x') = \dim H^1(\mathcal{A}+x+x')$

$$\begin{array}{ccccc} & & H^1(\mathcal{A}) & & \\ & \subseteq & & \supseteq & \\ H^1(\mathcal{A}+x) & & & & H^1(\mathcal{A}+x') \\ & \supseteq & & \subseteq & \\ & & H^1(\mathcal{A}+x+x') & & \end{array}$$

and likewise with x and x' interchanged, the fact that $H^1(\mathcal{A})/H^1(\mathcal{A}+x+x')$ is one-dimensional implies that *either* $H^1(\mathcal{A}+x) = H^1(\mathcal{A})$ and $H^1(\mathcal{A}+x') = H^1(\mathcal{A}+x+x')$ *or* $H^1(\mathcal{A}+x') = H^1(\mathcal{A})$ and $H^1(\mathcal{A}+x) = H^1(\mathcal{A}+x+x')$, which proves the assertion. The coset of P giving rise to $\mathcal{B}$ with $i(\mathcal{B})$ odd is $P+\delta_1 = P + \left\{\begin{matrix} \frac12 & 0 & \dots & 0 \\ 0 & 0 & \dots & 0\end{matrix}\right\}_{\hat{\tau}}$ since $\hat{\theta}\begin{bmatrix} \frac12 & 0 & \dots & 0 \\ \frac12 & 0 & \dots & 0\end{bmatrix}(\hat{z}) \equiv 0$ for $\hat{z} \in P$ by (64), (66), (69) and the fact that $\lambda+\delta_1 = \left\{\begin{matrix} \frac12 & 0 & \dots & 0 \\ \frac12 & 0 & \dots & 0\end{matrix}\right\}_{\hat{\tau}}$ is *odd*.

Proposition 4.11. For all $s \in \mathbb{C}^{g-1}$, the universal cover of P_0:

$$\hat{\theta}[\lambda](\phi(s)) = \frac{1}{c}\eta^2(s) \tag{78}$$

where c is the constant of Prop. 4.1 and ϕ is the isomorphism (70) used in the construction of the Prym variety. If $z \in \mathbb{C}^g$ and $s \in \mathbb{C}^{g-1}$:

$$\frac{\hat{\theta}\begin{bmatrix} 0 & \alpha & \alpha \\ \frac12 & \beta & \beta\end{bmatrix}(\pi^* z)}{\theta\begin{bmatrix} 0 & \alpha \\ 0 & \beta\end{bmatrix}(z)\,\theta\begin{bmatrix} 0 & \alpha \\ \frac12 & \beta\end{bmatrix}(z)} = c = (-1)^{4\gamma\cdot\delta}\,\frac{\eta^2\begin{bmatrix}\gamma \\ \delta\end{bmatrix}(s)}{\hat{\theta}\begin{bmatrix} 0 & \gamma & \gamma \\ \frac12 & \delta & \delta\end{bmatrix}(\phi(s))} \tag{79}$$

and

$$c = \frac{\eta_{2\pi}\begin{bmatrix}\rho \\ 0\end{bmatrix}(0)}{\theta_{2\tau}\begin{bmatrix} 0 & \rho \\ \frac12 & 0\end{bmatrix}(0)}, \qquad c^2 = \frac{\eta^2\begin{bmatrix}\alpha \\ \beta\end{bmatrix}(0)}{\theta\begin{bmatrix} 0 & \alpha \\ 0 & \beta\end{bmatrix}(0)\,\theta\begin{bmatrix} 0 & \alpha \\ \frac12 & \beta\end{bmatrix}(0)} \quad * \tag{80}$$

for all half-integer characteristics α, β, γ, δ and ρ.

---

* These formulas have been proved by Farkas-Rauch in [10].

Proof. (79) and the second term of (80) come from (67) and (78). To establish (78) and the first term of (80), we will use (75) and (76) of Prop. 4.8 which for $p = 2$ become

$$(75)' \qquad \hat{\Theta}_{\hat{T}}\begin{bmatrix} a_0 & a & c \\ b_0 & b & d \end{bmatrix}(\hat{z}) = \sum_{2\delta\in(\mathbb{Z}/2\mathbb{Z})^{g-1}} \Theta_{2T}\begin{bmatrix} a_0 & \frac{a+c}{2}+\delta \\ b_0 & b+d \end{bmatrix}(z)\,\eta_{2\pi}\begin{bmatrix} \frac{a-c}{2}+\delta \\ b-d \end{bmatrix}(s)$$

and

$$(76)' \quad 2^{g-1}\,\Theta_T\begin{bmatrix} a_0 & a \\ b_0 & b \end{bmatrix}(z)\,\eta_\pi\begin{bmatrix} c \\ d \end{bmatrix}(s) = \sum_{2\varepsilon\in(\mathbb{Z}/2\mathbb{Z})^{g-1}} e^{-4\pi i\varepsilon\cdot c}\,\hat{\Theta}_{\frac{\hat{T}}{2}}\begin{bmatrix} a_0 & a+c & a-c \\ b_0 & \frac{b+d}{2}+\varepsilon & \frac{b-d}{2}-\varepsilon \end{bmatrix}(\hat{z})$$

for all $a_0, b_0 \in \mathbb{R}$ and $a,b,c,d \in \mathbb{R}^{g-1}$. Now apply the transformation law of theta functions (3) to (67) and (76)'; then for $2\rho \in \mathbb{Z}^{g-1}$,

$$2^{g-1}\,\Theta_{2T}\begin{bmatrix} 0 & \rho \\ \frac{1}{2} & 0 \end{bmatrix}(0)\,\eta_{2\pi}\begin{bmatrix} \rho \\ 0 \end{bmatrix}(0) = \sum_{2\varepsilon\in(\mathbb{Z}/2\mathbb{Z})^{g-1}} (-1)^{4\rho\cdot\varepsilon}\,\hat{\Theta}_{\hat{T}}\begin{bmatrix} 0 & 0 & 0 \\ \frac{1}{2} & \varepsilon & \varepsilon \end{bmatrix}(0)$$

$$= c\sum_{2\varepsilon\in(\mathbb{Z}/2\mathbb{Z})^{g-1}} (-1)^{4\rho\cdot\varepsilon}\,\Theta_T\begin{bmatrix} 0 & 0 \\ 0 & \varepsilon \end{bmatrix}(0)\,\Theta_T\begin{bmatrix} 0 & 0 \\ \frac{1}{2} & \varepsilon \end{bmatrix}(0) = 2^{g-1}c\,\Theta^2_{2T}\begin{bmatrix} 0 & \rho \\ \frac{1}{2} & 0 \end{bmatrix}(0)$$

so $\eta_{2\pi}\begin{bmatrix} \rho \\ 0 \end{bmatrix}(0) = c\,\Theta_{2\tau}\begin{bmatrix} 0 & \rho \\ \frac{1}{2} & 0 \end{bmatrix}(0)$. Using (4), (80) and (75)', we get (78):

$$\hat{\Theta}_{\hat{T}}\begin{bmatrix} 0 & 0 & 0 \\ \frac{1}{2} & 0 & 0 \end{bmatrix}(\phi(s)) = \sum_{2\delta\in(\mathbb{Z}/2\mathbb{Z})^{g-1}} \Theta_{2T}\begin{bmatrix} 0 & \delta \\ \frac{1}{2} & 0 \end{bmatrix}(0)\,\eta_{2\pi}\begin{bmatrix} \delta \\ 0 \end{bmatrix}(2s)$$

$$= \frac{1}{c}\sum_{2\delta\in(\mathbb{Z}/2\mathbb{Z})^{g-1}} \eta_{2\pi}\begin{bmatrix} \delta \\ 0 \end{bmatrix}(0)\,\eta_{2\pi}\begin{bmatrix} \delta \\ 0 \end{bmatrix}(2s) = \frac{1}{c}\,\eta^2(s)$$

Corollary 4.12. If $\operatorname{div}_{\hat{C}}\eta(\int_x^y w - e) = \zeta \neq \hat{C}$ for $e \in P_0$ and fixed $x \in \hat{C}$, then $\phi(e) = \zeta + x' - x - \pi^*\Delta \in P$ with $i(\zeta) = 1$. Equivalently, if $\operatorname{div}_{\hat{C}}\eta(\frac{1}{2}\int_{y'}^{y} w - e) = \mathcal{A} \neq \hat{C}$, then $\phi(e) + \delta_1 = \mathcal{A} - \pi^*\Delta \in P + \delta_1$ with $i(\mathcal{A}) = 1$ and δ_1 the half period $\begin{Bmatrix} \frac{1}{2} & 0 & \cdots & 0 \\ 0 & 0 & \cdots & 0 \end{Bmatrix}_{\hat{T}} \in J(\hat{C})$.

<u>Proof</u>. If $e \in P_0$ is such that $\eta(\tfrac{1}{2}\int_x^{x'} w - e) \not\equiv 0$, then by Prop. 4.11, $\hat{\theta}\begin{bmatrix} \tfrac{1}{2} & 0 & \dots & 0 \\ \tfrac{1}{2} & 0 & \dots & 0 \end{bmatrix}(\int_x^y u - \phi(e)) \not\equiv 0$ on $\hat{C}$ since at $y = x'$,

$$\hat{\theta}[\lambda](\phi(\tfrac{1}{2}\int_x^{x'} w - e)) = \frac{1}{c}\eta^2(\tfrac{1}{2}\int_x^{x'} w - e) \neq 0.$$

Therefore $\mathrm{div}_{\hat{C}}\,\hat{\theta}\begin{bmatrix} \tfrac{1}{2} & 0 & \dots & 0 \\ \tfrac{1}{2} & 0 & \dots & 0 \end{bmatrix}(\int_x^y u - \phi(e)) = \mathcal{B} + x$ with $i(\mathcal{B}) = 1$ and $\mathcal{B} - \pi^*\Delta = \phi(e) + \delta_1 = \mathcal{A} - \pi^*\Delta$ by (77), which means that $\mathcal{A}$ is in fact the divisor $\mathcal{B}$ and $i(\mathcal{A}) = 1$ (see (88) below). The two statements in the corollary are equivalent since $\delta_1 + \phi(\tfrac{1}{2}\int_x^{x'} w) = x' - x \in P + \delta_1$ and $\int_x^y w = \tfrac{1}{2}\int_{y'}^y w + \tfrac{1}{2}\int_x^{x'} w \quad \forall\, x,y \in \hat{C}$.

From this corollary and Prop. 4.10 we have an analogue of Riemann's Theorem 1.1 for the η-function.

<u>Corollary 4.13</u> [34, p. 94]. Let

$$(\eta)_0 = P_0 - (\eta) = \{f \in P_0 \mid \eta(f) \neq 0\}$$

$$(\eta)_1 = \{f \in P_0 \mid \eta(\tfrac{1}{2}\int_x^{x'} w - f) \not\equiv 0 \text{ on } \hat{C}\}$$

and for $2 \leq n \leq g$,

$$(\eta)_n = \{f \in P_0 \mid \eta(\tfrac{1}{2}\int_{x_1+\dots+x_{n-2}}^{x_1'+\dots+x_{n-2}'} w - f) \equiv 0 \quad \forall\, x_1,\dots,x_{n-2} \in \hat{C} \text{ but}$$

$$\eta(\tfrac{1}{2}\int_{x_1+\dots+x_n}^{x_1'+\dots+x_n'} w - f) \not\equiv 0 \quad \forall\, x_1,\dots,x_n \in \hat{C}\}.$$

Then

$$(\eta)_n = \{f \in P_0 \mid \phi(f) + n\delta_1 = \mathcal{B} - \pi^*\Delta,\ \mathcal{B} \text{ a positive divisor of degree } \hat{g}\text{-}1 \text{ on } \hat{C} \text{ with } \pi_*(\mathcal{B}) = K_C \text{ and } i(\mathcal{B}) = n\}.$$

Also $(\eta) = \bigsqcup_{m\geq 1} (\eta)_{2m}$ and $P_0 = \bigsqcup_{m\geq 0} (\eta)_{2m+1}$, and (η) has multiplicity at least m at $f \in (\eta)_{2m}$ since

$$(\eta)_{2m} = \{f \in P_0 \mid \eta\Big(\int_{x_1+\dots+x_{m-1}}^{y_1+\dots+y_{m-1}} w - f\Big) \equiv 0 \quad \forall x_1,\dots,x_{m-1},y_1,\dots,y_{m-1} \in \hat{C}$$

$$\text{but } \eta\Big(\int_{x_1+\dots+x_m}^{y_1+\dots+y_m} w - f\Big) \not\equiv 0 \quad \forall x_1,\dots,x_m,y_1,\dots,y_m \in \hat{C}\}.$$

This corollary appears in the work of Wirtinger [34] who studied extensively the case $g = 4$: then a 3-dimensional Prym variety P is generically the Jacobian of a non-singular quartic curve, and as s_0 varies over P_0, $\operatorname{div}_{P_0}\eta(s-s_0) \cap \operatorname{div}_{P_0}\eta(s+s_0)$ defines a 3-dimensional family of Riemann surfaces $\hat{C}_{s_0}$ of genus 7 with Prym variety P, conformal involution $s \to -s + \tfrac{1}{2}s_0$, and quotient Riemann surface C_{s_0} of genus 4 a Wirtinger sextic curve constructed from the quartic curve. A discussion of this relation between the genus 3 and genus 4 Riemann surfaces can be found in [8] and in the thesis of Recillas [26].

<u>Proposition 4.14</u> (Riemann). If $e \in \mathbb{C}^{g-1}$ and c is the constant of Prop. 4.1,

$$\frac{\hat{\Theta}[\lambda]\big(\int_x^y u - \phi(e)\big)}{\eta(e)\eta\big(\int_x^y w - e\big)} = \frac{E(x,y)}{c\hat{E}(x,y)} \tag{81}$$

and

$$\frac{\hat{\Theta}[\lambda]\big(\int_x^y u - \phi(e) + \delta_1\big)}{\eta\big(\tfrac{1}{2}\int_x^{x'} w - e\big)\eta\big(\tfrac{1}{2}\int_y^{y'} w + e\big)} = \frac{E(x,y')}{c\hat{E}(x,y')} . \tag{82}$$

<u>Proof</u>. (82) follows from (81) by replacing y with y' and e by $e + \tfrac{1}{2}\int_y^{y'} w$. To prove (81): if (for generic e) $\operatorname{div}_{\hat{C}}\eta\big(\int_x^y w - e\big) = \zeta$ and $\operatorname{div}_{\hat{C}}\hat{\Theta}[\lambda]\big(\int_x^y u - \phi(e)\big) = \mathcal{A}$ then by (77),

$$\mathcal{A} - x - \pi^*\Delta = \phi(e) = \zeta + x' - x - \pi^*\Delta$$

so that $\mathcal{A}$ is actually the divisor $\zeta + x'$ since $i(\mathcal{A}) = 0$. Thus

$$\frac{\hat{E}(x,y)}{E(x,y)} \frac{\hat{\theta}[\lambda](\int_x^y u - \phi(e))}{\eta(\int_x^y w - e)}$$

has no zeroes or poles on $\hat{C}$ and is a meromorphic function by (72)' and Corollary 4.2; letting $y \to x$, this constant is $\frac{1}{c}\eta(e)$ by Prop. 4.11.

Corollary 4.15. For all $x,y \in \hat{C}$ and $2\alpha, 2\beta, 2\rho \in \mathbb{Z}^{g-1}$:

$$\text{(83)} \qquad \frac{\eta^2\begin{bmatrix}\alpha\\ \beta\end{bmatrix}(\tfrac{1}{2}\int_x^y w)}{\theta\begin{bmatrix}0 & \alpha\\ 0 & \beta\end{bmatrix}(\tfrac{1}{2}\int_x^y v)\,\theta\begin{bmatrix}0 & \alpha\\ \tfrac{1}{2} & \beta\end{bmatrix}(\tfrac{1}{2}\int_x^y v)} = c^2\,\frac{\hat{E}(x,y)}{E(x,y)}$$

and

$$\text{(84)} \qquad \frac{\eta_{2\pi}\begin{bmatrix}\rho\\ 0\end{bmatrix}(\int_x^y w)}{\theta_{2\tau}\begin{bmatrix}0 & \rho\\ \tfrac{1}{2} & 0\end{bmatrix}(\int_x^y v)} = c\,\frac{\hat{E}(x,y)}{E(x,y)}\,.$$

Proof. (83), (80) and (3) give (84). For (83), set $e = \tfrac{1}{2}\int_x^y w + \begin{Bmatrix}\alpha\\ \beta\end{Bmatrix}_\pi$ in (81) and use (67).

Taking a Taylor expansion of (83) near $y = x$, we get

Corollary 4.16. Let $\begin{bmatrix}\alpha\\ \beta\end{bmatrix}$ be an even half period, and set

$$R\begin{bmatrix}\alpha\\ \beta\end{bmatrix}(x) = \sum_{i,j=1}^{g-1} \frac{\partial^2 \ln \eta\begin{bmatrix}\alpha\\ \beta\end{bmatrix}}{\partial s_i\,\partial s_j}(0)\,w_i(x)\,w_j(x) \quad \text{and} \quad Q\begin{bmatrix}0 & \alpha\\ v & \beta\end{bmatrix}(x) = \sum_{i,j=0}^{g-1} \frac{\partial^2 \ln \theta\begin{bmatrix}0 & \alpha\\ v & \beta\end{bmatrix}}{\partial z_i\,\partial z_j}(0)\,v_i(x)\,v_j(x)$$

Then if $\hat{\omega}$ is the differential of the second kind on $\hat{C} \times \hat{C}$,

$$2R\begin{bmatrix}\alpha\\ \beta\end{bmatrix}(x) = 4\hat{\omega}(x,x') + Q\begin{bmatrix}0 & \alpha\\ 0 & \beta\end{bmatrix}(x) + Q\begin{bmatrix}0 & \alpha\\ \tfrac{1}{2} & \beta\end{bmatrix}(x) \qquad \forall x \in \hat{C}.$$

Proof. Use (83), (26) and the relation $6\hat{\omega}(x,\phi(x)) = S(x) - \hat{S}(x)$ from (61) and Cor. 2.6(i). This corollary can also be seen from (79) and the identity

$$\hat{\omega}(x,x') = -\sum_{i,j=0}^{2g-2} \frac{\partial^2 \ln \hat{\Theta}\begin{bmatrix}0 & \alpha & \alpha\\ \frac{1}{2} & \beta & \beta\end{bmatrix}}{\partial \hat{z}_i \partial \hat{z}_j}(0)u_i(x)u_j(x') \qquad \forall\, x \in \hat{C}$$

coming from (29) and the fact that $\hat{\Theta}\begin{bmatrix}0 & \alpha & \alpha\\ \frac{1}{2} & \beta & \beta\end{bmatrix}(\int_x^{x'} u) \equiv 0$ on $\hat{C}$ by Prop. 4.10.

Corollary 4.17. Let $\begin{bmatrix}\gamma\\ \delta\end{bmatrix}$ be an odd half-period, and set

$$G\begin{bmatrix}\gamma\\ \delta\end{bmatrix}(x) = \sum_{1}^{g-1} \frac{\partial \eta\begin{bmatrix}\gamma\\ \delta\end{bmatrix}}{\partial s_i}(0)w_i(x) \quad \text{and} \quad H\begin{bmatrix}0 & \gamma\\ \nu & \delta\end{bmatrix}(x) = \sum_{0}^{g-1} \frac{\partial \Theta\begin{bmatrix}0 & \gamma\\ \nu & \delta\end{bmatrix}}{\partial z_i}(0)v_i(x).$$

Then

$$G\begin{bmatrix}\gamma\\ \delta\end{bmatrix}^2(x) = c^2 H\begin{bmatrix}0 & \gamma\\ 0 & \delta\end{bmatrix}(x) H\begin{bmatrix}0 & \gamma\\ \frac{1}{2} & \delta\end{bmatrix}(x) \qquad \forall\, x \in \hat{C}. \tag{85}$$

The "Schottky-relations" (80) and (85), together with the Riemann theta-formula [17, p. 137] or [19, p. 308], can be used to give equations, respectively, of hypersurfaces in $\mathfrak{H}_g$ containing the locus of period matrices of Riemann surfaces of genus g, and of hypersurfaces in $\mathbb{P}_{g-1}(\mathbb{C})$ containing the image of C under the canonical mapping [13, p. 241], generalizing the Riemann-Weber irrational form of the equation of a non-hyperelliptic genus 3 Riemann surface as a non-singular quartic in $\mathbb{P}_2(\mathbb{C})$ [7, p. 387]. Both types of equation were first given in genus 4 by Schottky [28, p. 263 and p. 262 respectively] and later indicated for arbitrary genus in [29I, p. 296].

Proposition 4.18. For an odd characteristic $\begin{bmatrix}\gamma\\ \delta\end{bmatrix}$, let $f(\hat{\tau})$ be the function $\hat{\Theta}_{\hat{\tau}}\begin{bmatrix}0 & \gamma & \gamma\\ \frac{1}{2} & \delta & \delta\end{bmatrix}(0)$ on the Siegel half-plane $\mathfrak{H}_{2g-1}$ vanishing, by Prop. 4.11, on the locus $\hat{\mathcal{T}}_{2g-1} \subset \mathfrak{H}_{2g-1}$ of period matrices of curves $\hat{C}$ of genus $2g-1$ admitting a fixed-point-free involution. Then under the isomorphism (51) lifted to $\hat{\mathcal{T}}_{2g-1}$, the differential $df(\hat{\tau})$ restricted to $\hat{\mathcal{T}}_{2g-1}$ is the quadratic differential

$$cH\begin{bmatrix}0 & \gamma\\ 0 & \delta\end{bmatrix}(x)H\begin{bmatrix}0 & \gamma\\ \frac12 & \delta\end{bmatrix}(x) - \frac{1}{c}G\begin{bmatrix}\gamma\\ \delta\end{bmatrix}^2(x)$$

vanishing identically on $\hat{C}$ by (85).

<u>Proof</u>. If $e = \begin{Bmatrix}0 & \gamma & \gamma\\ \frac12 & \delta & \delta\end{Bmatrix}_{\hat{\mathcal{T}}} \in J(\hat{C})$ with $\begin{bmatrix}\gamma\\ \delta\end{bmatrix}$ odd, (79) gives

$$cH\begin{bmatrix}0 & \gamma\\ 0 & \delta\end{bmatrix}(x)H\begin{bmatrix}0 & \gamma\\ \frac12 & \delta\end{bmatrix}(x) = \sum_{i,j=0}^{2g-2} \frac{\partial^2\hat{\theta}[e]}{\partial\hat{z}_i\partial\hat{z}_j}(0)(\pi^*v(x))_i(\pi^*v(x))_j$$

and

$$-\frac{1}{c}G\begin{bmatrix}\gamma\\ \delta\end{bmatrix}^2(x) = \sum_{i,j=0}^{2g-2} \frac{\partial^2\hat{\theta}[e]}{\partial\hat{z}_i\partial\hat{z}_j}(0)(\phi w(x))_i(\phi w(x))_j .$$

From the symmetry (66) of $\hat{\theta}$, the universal covers $\tilde{P}$ and $\widetilde{\pi^*J}$ in $\mathbb{C}^{2g-1}$ are orthogonal under the quadratic form Q defined for any half period $e \in \pi^*J$ by $Q(\xi,\zeta) = \sum_{i,j=0}^{2g-2} \frac{\partial^2\hat{\theta}[e]}{\partial\hat{z}_i\partial\hat{z}_j}(0)\xi_i\zeta_j$, $\xi,\zeta \in \mathbb{C}^{\hat{g}}$. Therefore, the quadratic differential corresponding to df under the isomorphism (51) lifted to $\hat{\mathcal{T}}_{2g-1}$ is, by the heat equation (52),

$$\sum_{0\le i\le j\le 2g-2} \frac{\partial f}{\partial\hat{\tau}_{ij}}(u_i(x)u_j(x) + u_i(x')u_j(x')) = 4\sum_{i,j=0}^{2g-2} \left.\frac{\partial^2\hat{\theta}[e]}{\partial\hat{z}_i\partial\hat{z}_j}\right|_{z=0} u_i(x)u_j(x)$$

$$= cH\begin{bmatrix}0 & \gamma\\ 0 & \delta\end{bmatrix}(x)H\begin{bmatrix}0 & \gamma\\ \frac12 & \delta\end{bmatrix}(x) - \frac{1}{c}G\begin{bmatrix}\gamma\\ \delta\end{bmatrix}^2(x).$$

In similar manner, if $e = \begin{Bmatrix}0 & \alpha & \alpha\\ \frac12 & \beta & \beta\end{Bmatrix}_{\hat{\mathcal{T}}}$ with $\begin{bmatrix}\alpha\\ \beta\end{bmatrix}$ even, Prop. 4.5 and (26) give

$$\sum_{i,j=0}^{2g-2} \frac{\partial^2 \ln\hat{\theta}[e]}{\partial\hat{z}_i\partial\hat{z}_j}(0)u_i(x)u_j(x) = \hat{\omega}(x,x') + \tfrac12\left\{Q\begin{bmatrix}0 & \alpha\\ 0 & \beta\end{bmatrix}(x) + Q\begin{bmatrix}0 & \alpha\\ \frac12 & \beta\end{bmatrix}(x)\right\},$$

so that the quadratic differential of Cor. 4.16

$$2\hat{\theta}\begin{bmatrix}0 & \alpha & \alpha\\ \frac12 & \beta & \beta\end{bmatrix}^2(0)\left(4\hat{\omega}(x,x') + Q\begin{bmatrix}0 & \alpha\\ 0 & \beta\end{bmatrix}(x) + Q\begin{bmatrix}0 & \alpha\\ \frac12 & \beta\end{bmatrix}(x) - 2R\begin{bmatrix}\alpha\\ \beta\end{bmatrix}(x)\right),$$

which vanishes identically on $\hat{C}$, is the differential $dg|\hat{\mathcal{T}}_{2g-1}$ where

$$g(\hat{\mathcal{T}}) = \hat{\Theta}^2_{\hat{\mathcal{T}}}\begin{bmatrix} 0 & \alpha & \alpha \\ \frac{1}{2} & \beta & \beta \end{bmatrix}(0) - \frac{1}{4^{g-1}} \sum_{2\epsilon, 2\epsilon' \in \mathbb{Z}^{g-1}} (-1)^{4\alpha\cdot(\epsilon+\epsilon')} \hat{\Theta}_{\frac{\hat{\mathcal{T}}}{2}}\begin{bmatrix} 0 & 0 & 0 \\ 0 & \beta+\epsilon & \epsilon \end{bmatrix}(0)\, \hat{\Theta}_{\frac{\hat{\mathcal{T}}}{2}}\begin{bmatrix} 0 & 0 & 0 \\ \frac{1}{2} & \beta+\epsilon' & \epsilon' \end{bmatrix}(0)$$

vanishes identically on $\hat{\mathcal{T}}_{2g-1}$ by (76)' and (79).

<u>Proposition 4.19</u> (Schottky-Jung). If $\begin{bmatrix}\alpha\\ \beta\end{bmatrix}$ and $\begin{bmatrix}\gamma\\ \delta\end{bmatrix}$ are, respectively, even and odd (g-1)-characteristics,

$$(86) \qquad \frac{E^2(x,y)}{\hat{E}^2(x,y)} \frac{\eta\begin{bmatrix}\alpha\\ \beta\end{bmatrix}(\int_x^y w)}{\eta\begin{bmatrix}\alpha\\ \beta\end{bmatrix}(0)} = \frac{1}{2}\left\{ \frac{\Theta\begin{bmatrix}0 & \alpha\\ 0 & \beta\end{bmatrix}(\int_x^y v)}{\Theta\begin{bmatrix}0 & \alpha\\ 0 & \beta\end{bmatrix}(0)} + \frac{\Theta\begin{bmatrix}0 & \alpha\\ \frac{1}{2} & \beta\end{bmatrix}(\int_x^y v)}{\Theta\begin{bmatrix}0 & \alpha\\ \frac{1}{2} & \beta\end{bmatrix}(0)} \right\}$$

and

$$(87) \qquad \frac{E^2(x,y)}{\hat{E}^2(x,y)} \frac{\eta\begin{bmatrix}\gamma\\ \delta\end{bmatrix}(\int_x^y w)}{G\begin{bmatrix}\gamma\\ \delta\end{bmatrix}(x)} = \frac{1}{2}\left\{ \frac{\Theta\begin{bmatrix}0 & \gamma\\ 0 & \delta\end{bmatrix}(\int_x^y v)}{H\begin{bmatrix}0 & \gamma\\ 0 & \delta\end{bmatrix}(x)} + \frac{\Theta\begin{bmatrix}0 & \gamma\\ \frac{1}{2} & \delta\end{bmatrix}(\int_x^y v)}{H\begin{bmatrix}0 & \gamma\\ \frac{1}{2} & \delta\end{bmatrix}(x)} \right\}$$

for all $x,y \in \hat{C}$.

<u>Proof</u>. (86) comes from taking $e = \begin{Bmatrix}0 & \alpha\\ 0 & \beta\end{Bmatrix}_\tau \in J(C)$ in Prop. 4.5 and $e = \begin{Bmatrix}\alpha\\ \beta\end{Bmatrix}_\pi \in P_0$ in (81), using (79)-(80). To prove (87), take $e = \int_x^z v + \begin{Bmatrix}0 & \gamma\\ 0 & \delta\end{Bmatrix}_\tau \in J(C)$ in Prop. 4.5 and $e = \int_x^z w + \begin{Bmatrix}\gamma\\ \delta\end{Bmatrix}_\pi \in P_0$ in (81), and let $z \to x$, using (85).

From these identities we find that the differentials involved in Cor. 4.9 for $e \in P$ a half-period are, by (39) and (25):

$$4\,\frac{E(x,y)E(x,y')\eta\begin{bmatrix}\alpha\\ \beta\end{bmatrix}(\int_x^y w)\eta\begin{bmatrix}\alpha\\ \beta\end{bmatrix}(\int_x^{y'} w)}{\hat{E}^2(x,y)\hat{E}^2(x,y')\eta\begin{bmatrix}\alpha\\ \beta\end{bmatrix}^2(0)} = \sum_{i,j=1}^{g} \frac{\partial^2}{\partial z_i \partial z_j} \ln \frac{\Theta\begin{bmatrix}0 & \alpha\\ 0 & \beta\end{bmatrix}}{\Theta\begin{bmatrix}0 & \alpha\\ \frac{1}{2} & \beta\end{bmatrix}}(0) v_i(x) v_j(y)$$

and

$$4\,\frac{E(x,y)E(x,y')\eta\begin{bmatrix}\gamma\\ \delta\end{bmatrix}(\int_x^y w)\eta\begin{bmatrix}\gamma\\ \delta\end{bmatrix}(\int_x^{y'} w)}{\hat{E}^2(x,y)\hat{E}^2(x,y')G\begin{bmatrix}\gamma\\ \delta\end{bmatrix}^2(x)} = \frac{H\begin{bmatrix}0 & \alpha\\ 0 & \beta\end{bmatrix}(y)}{H\begin{bmatrix}0 & \alpha\\ 0 & \beta\end{bmatrix}(x)} - \frac{H\begin{bmatrix}0 & \alpha\\ \frac{1}{2} & \beta\end{bmatrix}(y)}{H\begin{bmatrix}0 & \alpha\\ \frac{1}{2} & \beta\end{bmatrix}(x)}$$

for all $x,y \in \hat{C}$.

As a consequence of Props. 4.14 and 4.19, one has:

Corollary 4.20. Let $\varepsilon(x)$ be the non-vanishing multiplicative holomorphic differential $\hat{E}(x,x')^{-1}$ on $\hat{C}$; then $\varepsilon(x') = \varepsilon(x) \in H^0(\pi^* K_C \otimes \nabla^*(\eta)^{-2})$, where $\nabla\colon \hat{C} \to P_0$ is the difference mapping $\nabla x = \tfrac{1}{2}\phi^{-1} \circ \sigma_2(x'-x)$. For all $x \in \hat{C}$,

$$\frac{\sum_0^{g-1} \frac{\partial \Theta_{2\tau}\begin{bmatrix} \frac{1}{2} & \rho \\ \frac{1}{2} & 0 \end{bmatrix}}{\partial z_i}(0) v_i(x)}{\eta_{2\pi}\begin{bmatrix} \rho \\ 0 \end{bmatrix}\left(\int_x^{x'} w\right)} = \frac{1}{c} e^{\frac{1}{2}\pi i - \frac{1}{4}\tau_{00}} \varepsilon(x), \qquad 2\rho \in \mathbf{Z}^{g-1}$$

$$(88) \qquad \frac{\sum_0^{2g-2} \frac{\partial \hat{\Theta}[\lambda]}{\partial \hat{z}_i}(\delta_1 - \phi(s)) u_i(x)}{\eta\left(\frac{1}{2}\int_x^{x'} w - s\right)\eta\left(\frac{1}{2}\int_x^{x'} w + s\right)} = \frac{\varepsilon(x)}{c e^{\frac{1}{2}\tau_{00}}} \qquad \forall s \in P_0$$

$$\frac{\eta\begin{bmatrix} \alpha \\ \beta \end{bmatrix}(0)\frac{d^2}{dy^2}\ln \left.\frac{\Theta\begin{bmatrix} 0 & \alpha \\ \frac{1}{2} & \beta \end{bmatrix}\left(\int_x^y v\right)}{\Theta\begin{bmatrix} 0 & \alpha \\ 0 & \beta \end{bmatrix}\left(\int_x^y v\right)}\right|_{y=x}}{\eta\begin{bmatrix} \alpha \\ \beta \end{bmatrix}\left(\int_x^{x'} w\right)} = \frac{G\begin{bmatrix} \gamma \\ \delta \end{bmatrix}(x)\frac{d}{dx}\ln \frac{H\begin{bmatrix} 0 & \gamma \\ \frac{1}{2} & \delta \end{bmatrix}(x)}{H\begin{bmatrix} 0 & \gamma \\ 0 & \delta \end{bmatrix}(x)}}{\eta\begin{bmatrix} \gamma \\ \delta \end{bmatrix}\left(\int_x^{x'} w\right)} = -4e^{-\frac{1}{2}\tau_{00}} \varepsilon^2(x)$$

and

$$(89) \qquad \frac{\Theta\begin{bmatrix} \frac{1}{2} & \alpha \\ 0 & \beta \end{bmatrix}(0) H\begin{bmatrix} \frac{1}{2} & \alpha \\ \frac{1}{2} & \beta \end{bmatrix}(x)}{\eta^2\begin{bmatrix} \alpha \\ \beta \end{bmatrix}\left(\frac{1}{2}\int_x^{x'} w\right)} = \frac{\Theta\begin{bmatrix} \frac{1}{2} & \gamma \\ \frac{1}{2} & \delta \end{bmatrix}(0) H\begin{bmatrix} \frac{1}{2} & \gamma \\ 0 & \delta \end{bmatrix}(x)}{\eta^2\begin{bmatrix} \gamma \\ \delta \end{bmatrix}\left(\frac{1}{2}\int_x^{x'} w\right)} = \frac{2}{c^2} e^{\frac{1}{2}\pi i - \frac{1}{4}\tau_{00}} \varepsilon(x)$$

for all even $\begin{bmatrix} \alpha \\ \beta \end{bmatrix}$ and odd $\begin{bmatrix} \gamma \\ \delta \end{bmatrix}$.

Proof. Let $y \to x'$ in (84), $y \to x$ in (82), $y \to x$ in (86)-(87) and $y \to x'$ in (83) making use of the fact that, as a multiplicative section of $\pi^* L_0^{-1}$ on $\hat{C}$,

$$E(x,y') = E(x,y)\exp\left\{-\tfrac{1}{2}\tau_{00} - \int_x^y v_0\right\} \qquad \forall\, x,y \in \hat{C}.$$

In all these formulas the path of integration from x to x' is kept within $\hat{C}$ dissected along the homology base defining $\hat{\tau}$ and the sections $\hat{E}$ and ε.

If $K(P_0)$ is the quotient of P_0 by the involution $s \to -s$, then $K(P_0) \hookrightarrow \mathbb{P}_{2^{g-1}-1}(\mathbb{C})$ by means of the 2^{g-1} linearly independent second-order η-functions of characteristics $\begin{bmatrix}0\\0\end{bmatrix}$ on P_0 [9, pp. 212-220]; it will now be shown that the composite mapping $\hat{C} \xrightarrow{\nabla} P_0 \longrightarrow K(P_0) \longrightarrow \mathbb{P}_{2^{g-1}-1}$ actually projects to a mapping $C \to \mathbb{P}_{g-1} \longrightarrow \mathbb{P}_{2^{g-1}-1}$ which is the canonical imbedding for non-hyperelliptic C.

<u>Corollary 4.21</u> (Mumford). For $\begin{bmatrix}\alpha\\\beta\end{bmatrix}$ even and $\begin{bmatrix}\gamma\\\delta\end{bmatrix}$ odd, the 4^{g-1} differentials $H\begin{bmatrix}\frac{1}{2} & \alpha\\\frac{1}{2} & \beta\end{bmatrix}(x)$ and $H\begin{bmatrix}\frac{1}{2} & \gamma\\0 & \delta\end{bmatrix}(x)$ span $H^0(C,\Omega_C^1)$.

<u>Proof</u>. The linear series $\pi^*|K_C|$ and $\nabla^*|2\eta|_{P_0}$ are the same since as e varies over P_0, $\mathrm{div}_{\hat{C}}\,\eta(\frac{1}{2}\int_x^{x'} w - e)\eta(\frac{1}{2}\int_x^{x'} w + e)$ is, by Cor. 4.13, either all of $\hat{C}$ or all divisors $\mathcal{A} + \mathcal{A}'$ on $\hat{C}$ with $i_{\hat{C}}(\mathcal{A}) = 1$ and $\pi_*(\mathcal{A}) = K_C$. Now use (89) and a theorem of Mumford [23, p. 297] which says that on any Abelian variety A, the linear series $|2\Theta|$ is generated by the squares of the odd and even theta-functions on A.

It should be noted for the purposes of the remark on p. 16 that this corollary does not require the precise statement (89) involving $\varepsilon(x)$ which itself was constructed from the prime form: all that is needed is

$$\mathrm{div}_{\hat{C}}\eta^2\begin{bmatrix}\alpha\\\beta\end{bmatrix}(\tfrac{1}{2}\int_x^{x'} w) = \mathrm{div}_{\hat{C}}H\begin{bmatrix}\frac{1}{2} & \alpha\\\frac{1}{2} & \beta\end{bmatrix}(x) \quad \text{and} \quad \mathrm{div}_{\hat{C}}\eta^2\begin{bmatrix}\gamma\\\delta\end{bmatrix}(\tfrac{1}{2}\int_x^{x'} w) = \mathrm{div}_{\hat{C}}H\begin{bmatrix}\frac{1}{2} & \gamma\\0 & \delta\end{bmatrix}(x)$$

which follows from Corollary 4.12 since $\eta^2[e](\frac{1}{2}\int_x^{x'} w)$, for e a half period, vanishes at the zeroes of a differential $H[e'](x)$, e' an odd half period satisfying $\pi^*(e') = \phi(e) + \delta_1$.

V. Ramified Double Coverings

This chapter is a discussion of theta-functions on Riemann surfaces $\hat{C}$ admitting a conformal involution with fixed points, the simplest class of surfaces $\hat{C}$ with non-trivial automorphism group Aut $\hat{C}$ and ramified projection mapping $\hat{C} \to \hat{C}/\text{Aut}\,\hat{C}$. Depending on the number of fixed points, the theory includes both the hyperelliptic theta-functions and the ramified double coverings with only two branch points, where the Θ-η relations become, in limiting cases, the Schottky relations (80) and (85) for unramified double coverings.

Relations between $\hat{\Theta}$ and Θ-functions. Let $\hat{C} \overset{\pi}{\to} C$ be a ramified double covering of genus $\hat{g} = 2g+n-1$ of a compact Riemann surface C of genus g with $2n$ branch points at $Q_1,\dots,Q_{2n} \in C$. If $\phi\colon \hat{C} \to \hat{C}$ is the conformal automorphism with fixed points at $Q_1,\dots,Q_{2n}$, a canonical homology basis of $H_1(\hat{C},\mathbb{Z})$

$$A_1,B_1,\dots,A_g,B_g,A_{g+1},B_{g+1},\dots,A_{g+n-1},B_{g+n-1},A_{1'},B_{1'},\dots,A_{g'},B_{g'}$$

can be chosen such that $A_1,B_1,\dots,A_g,B_g$ is a canonical basis of $H_1(C,\mathbb{Z})$ and

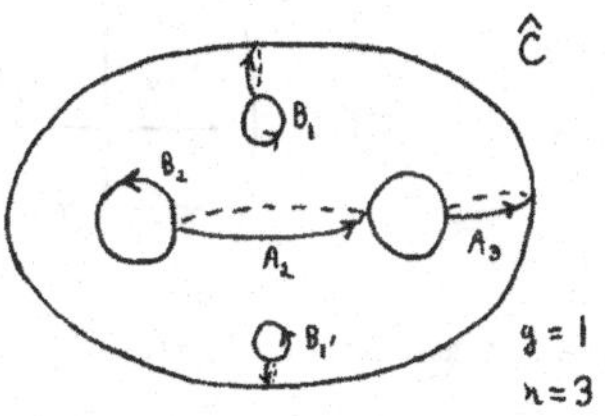

$$A_{\alpha'} + \phi(A_\alpha) = B_{\alpha'} + \phi(B_\alpha) = 0, \quad 1 \le \alpha \le g$$

$$A_i + \phi(A_i) = B_i + \phi(B_i) = 0, \quad g+1 \le i \le g+n-1.$$

If the corresponding normalized holomorphic differentials are

$$u_1,\dots,u_g,u_{g+1},\dots,u_{g+n-1},u_{1'},\dots,u_{g'}$$

then for $1 \le \alpha \le g$ and $g+1 \le i \le g+n-1$,

$$(90) \qquad u_\alpha(x) = -u_{\alpha'}(x') \quad \text{and} \quad u_i(x) = -u_i(x') \qquad \forall\, x \in \hat{C}$$

where $x' = \phi(x)$ is the conjugate point of $x \in \hat{C}$. The normalized holomorphic differentials on C are then given by $v_\alpha = u_\alpha - u_{\alpha'}$ for $1 \le \alpha \le g$, while

$$w_\alpha = u_\alpha + u_{\alpha'}, \quad 1 \le \alpha \le g \quad \text{and} \quad w_i = u_i \quad g+1 \le i \le g+n-1$$

are $g+n-1$ linearly independent normalized Prym differentials on $\hat{C}$. The canonical bilinear differential and prime form for $\hat{C}$ have the symmetries

$$\hat{\omega}(x,y) = \hat{\omega}(x',y') \quad \text{and} \quad \hat{E}^2(x,y) = \hat{E}^2(x',y') \qquad \forall\, x,y \in \hat{C}$$

and $\hat{\omega}(x,y) + \hat{\omega}(x,y')$ is the bilinear differential $\omega(x,y)$ on C. The Riemann matrix for $\hat{C}$ has the form

$$\text{(91)} \qquad \hat{T} = \left(\begin{array}{c|c|c} \frac{\pi_{\alpha\beta}+\tau_{\alpha\beta}}{2} & \pi_{\alpha j} & \frac{\pi_{\alpha\beta}-\tau_{\alpha\beta}}{2} \\ \hline \pi_{i\beta} & 2\pi_{ij} & \pi_{\alpha j}^{\,t} \\ \hline \frac{\pi_{\alpha\beta}-\tau_{\alpha\beta}}{2} & \pi_{i\beta}^{\,t} & \frac{\pi_{\alpha\beta}+\tau_{\alpha\beta}}{2} \end{array}\right) \qquad \begin{array}{l} 1 \le \alpha,\beta \le g \\ g+1 \le i,j \le g+n-1 \end{array}$$

where τ is the Riemann matrix for C and

$$\text{(92)} \qquad \Pi = \left(\begin{array}{c|c} \pi_{\alpha\beta} & \pi_{\alpha j} \\ \hline \pi_{i\beta} & \pi_{ij} \end{array}\right) = \left(\begin{array}{c|c} \int_{B_\beta} w_\alpha & \frac{1}{2}\int_{B_j} w_\alpha \\ \hline \int_{B_\beta} w_i & \frac{1}{2}\int_{B_j} w_i \end{array}\right) \qquad \alpha,\beta,i,j \text{ as above}$$

is a symmetric $(g+n-1)\times(g+n-1)$ matrix with negative-definite real part (see p. 95).

The field of meromorphic functions on $\hat{C}$ is obtained by adjoining to the functions on C the square root $\phi(x)$ of a function on C with simple zeroes or poles at $Q_1,\dots,Q_{2n}$ and double zeroes and poles elsewhere - for instance, $\phi^2(x) = \frac{\Theta(x-a-e-\rho)^2}{\Theta(x-a-e)^2} \prod_1^n \frac{E(x,Q_{i_k})}{E(x,Q_{j_k})}$ with $a \in C$, $e \in \mathbb{C}^g$, $\{i_1,\dots,i_n\} \cup \{j_1,\dots,j_n\}$ a partition of $\{1,\dots,2n\}$ and $\rho = \frac{1}{2}\sum_1^n \int_{Q_{i_k}}^{Q_{j_k}} v$. The map $\pi^*\colon J_0(C) \to J_0(\hat{C})$ lifting divisor classes of degree 0 is an injection: otherwise if $\pi^* \mathcal{A} = \operatorname{div}_{\hat{C}}(f+g\phi)$ with $\mathcal{A} \in J_0(C)$ and f,g meromorphic on C, the invariance of $\pi^*\mathcal{A}$ under ϕ implies $f \equiv 0$ or $g \equiv 0$ on C, which means g must actually vanish on C since $\operatorname{div}_{\hat{C}} \phi$ is not of form $\mathcal{B} + \phi(\mathcal{B})$ for any divisor $\mathcal{B}$. With the above choice of period matrix for $\hat{C}$, we can lift π^* to a mapping $\pi^*: \mathbb{C}^g \to \mathbb{C}^{\hat{g}}$ by:

$$\pi^* z = \pi^*(z_1,\ldots,z_g) = (z_1,\ldots,z_g,0,\ldots,0,-z_1,\ldots,-z_g) \in \mathbb{C}^{\hat{g}}$$

which, in characteristic notation, becomes

$$(93) \qquad \pi^* \begin{Bmatrix} \alpha \\ \beta \end{Bmatrix}_\tau = \begin{Bmatrix} \alpha & 0 & -\alpha \\ \beta & 0 & -\beta \end{Bmatrix}_{\hat{\tau}} \in \mathbb{C}^g \qquad \text{for } \alpha,\beta \in \mathbb{R}^g.$$

Similarly the automorphism $\phi: J_0(\hat{C}) \to J_0(\hat{C})$ arising from the involution ϕ on $\hat{C}$ can by (90) be lifted to an automorphism of $\mathbb{C}^{\hat{g}}$ sending $\hat{z} = (\hat{z}_1,\ldots,\hat{z}_{g+1},\ldots,\hat{z}_{1'},\ldots,\hat{z}_{g'})$ to

$$\phi(\hat{z}) = -(\hat{z}_{1'},\ldots,\hat{z}_{g'},\hat{z}_{g+1},\ldots,\hat{z}_{g+n-1},\hat{z}_1,\ldots,\hat{z}_g) \in \mathbb{C}^{\hat{g}}$$

that is,

$$(94) \quad \phi \begin{Bmatrix} \alpha & \mu & \delta \\ \beta & \nu & \varepsilon \end{Bmatrix}_{\hat{\tau}} = - \begin{Bmatrix} \delta & \mu & \alpha \\ \varepsilon & \nu & \beta \end{Bmatrix}_{\hat{\tau}} \in \mathbb{C}^{\hat{g}} \quad \text{for } \alpha,\beta \in \mathbb{R}^g, \quad \mu,\nu \in \mathbb{R}^{n-1}.$$

The theta function $\hat{\theta}$ constructed from $\hat{\tau}$ (91) has the symmetry

$$(95) \qquad \hat{\theta}(\hat{z}) = \hat{\theta}(\phi(\hat{z})) \qquad \forall\, \hat{z} \in \mathbb{C}^{\hat{g}}$$

which implies, by Riemann's Theorem 1.1 (10), that $\hat{\Delta} = \phi(\hat{\Delta})$ in $J_{\hat{g}-1}(\hat{C})$ for the divisor class $\hat{\Delta}$ corresponding to $\hat{\theta}$. If we let $D = \hat{\Delta} - \pi^*\Delta \in J_n(\hat{C})$ be the divisor class of degree n corresponding to the line bundle $\hat{L}_0 \otimes \pi^* L_0^{-1}$ on $\hat{C}$, then D is fixed under ϕ since $\hat{\Delta}$ is, and

$$(96) \qquad 2D = 2\hat{\Delta} - 2\pi^*\Delta = K_{\hat{C}} - \pi^* K_C = Q_1 + \ldots + Q_{2n} \in J_{2n}(\hat{C})$$

since the pullback by $d\pi$ of any holomorphic differential on C must vanish at $Q_1,\ldots,Q_{2n} \in \hat{C}$.

Proposition 5.1. For $z \in \mathbb{C}^g$ and $x_1,\ldots,x_n \in \hat{C}$,

$$\frac{\hat{\Theta}(\pi^* z + x_1 + \ldots + x_n - D)}{\Theta(z)\Theta(z + \pi_*(x_1 + \ldots + x_n - D))} = c(x_1,\ldots,x_n) \tag{97}$$

where $c(x_1,\ldots,x_n)$ is a holomorphic section of a line bundle on $\hat{C} \times \hat{C} \times \ldots \times \hat{C}$ which is symmetric in $x_1,\ldots,x_n$ and independent of $z \in \mathbb{C}^g$. Thus for $\delta, \varepsilon \in \mathbb{R}^g$:

$$\frac{\hat{\Theta}\begin{bmatrix}\delta & 0 & -\delta\\ \varepsilon & 0 & -\varepsilon\end{bmatrix}\left(z + \tfrac{1}{2}\int_D^{x_1+\ldots+x_n} w\ ,\ \int_D^{x_1+\ldots+x_n} w\ ,\ -z + \tfrac{1}{2}\int_D^{x_1+\ldots+x_n} w\right)}{\Theta\begin{bmatrix}\delta\\ \varepsilon\end{bmatrix}\left(z + \tfrac{1}{2}\int_D^{x_1+\ldots+x_n} v\right)\Theta\begin{bmatrix}\delta\\ \varepsilon\end{bmatrix}\left(z - \tfrac{1}{2}\int_D^{x_1+\ldots+x_n} v\right)} = c(x_1,\ldots,x_n). \tag{98}$$

Proof. Replacing z by $z - \tfrac{1}{2}\int_D^{x_1+\ldots+x_n} v$ in (97) gives (98). To prove (97), suppose $z = \zeta - \Delta \in (\Theta)$ with ζ positive of degree $g-1$ on C; then by definition of D,

$$\pi^* z + x_1 + \ldots + x_n - D = \pi^* \zeta + x_1 + \ldots + x_n - \hat{\Delta} \in (\hat{\Theta}).$$

Similarly, if $z + \pi_*(x_1 + \ldots + x_n - D) = \Delta - \xi \in (\Theta)$ with ξ positive of degree $g-1$ on C,

$$-\phi(\pi^* z + x_1 + \ldots + x_n - D) = \pi^*(-z) - x_1' - \ldots - x_n' + D' = \pi^* \xi + x_1 + \ldots + x_n - \hat{\Delta} \in (\hat{\Theta})$$

so that by the symmetry (95) of $\hat{\Theta}$, $\hat{\Theta}(\pi^* z + x_1 + \ldots + x_n - D) = 0$ again. Hence the left hand side of (97) is a well-defined meromorphic function on J(C) with poles at most on the $g-2$ dimensional subvariety $(\Theta) \cap (\Theta) + \pi_*(D - x_1 - \ldots - x_n)$, and so must be a constant $c(x_1,\ldots,x_n)$ finite for all $x_1,\ldots,x_n \in \hat{C}$. The section c is non-zero for generic $x_1,\ldots,x_n \in \hat{C}$ since for positive ζ of degree $g-1$ on C with $i_C(\zeta) = 1$ and for generic $x_1,\ldots,x_n$, $\zeta - \Delta + \pi_*(x_1 + \ldots + x_n - D) \notin (\Theta)$ and $i_{\hat{C}}(\pi^* \zeta + x_1 + \ldots + x_n) = 1$: otherwise if $i_C(\zeta) = 1$ and

$i_{\hat{C}}(\pi^*\zeta+x_1+\ldots+x_n) = d > 1$, there would be at least $d-1 > 0$ linearly independent Prym differentials vanishing at the generic divisor $\zeta + \pi_*(x_1+\ldots+x_n)$ of degree $g+n-1$ on C, contradicting the fact that there are only $g+n-1$ linearly independent Prym differentials on $\hat{C}$.

In the case $C = \mathbb{P}_1(\mathbb{C})$, $\hat{C}$ is hyperelliptic of genus $n-1$, $D = \hat{\Delta} - \pi^*\Delta = \hat{\Delta} + \pi^*p$ for any $p \in \mathbb{P}_1(\mathbb{C})$, and $c(x_1,\ldots,x_n)$ is the hyperelliptic theta function $\hat{\theta}(x_1+\ldots+x_n-\pi^*p-\hat{\Delta})$ which vanishes exactly when $x_i = x_j'$ for some $i \neq j$ (see p. 13). Similarly for genus $g > 0$, one has

<u>Corollary 5.2</u>. For $x,x_1,\ldots,x_{n-1} \in \hat{C}$ and $z \in \mathbb{C}^g$,

$$\frac{\hat{\theta}\left(\int_x^y u - \int_D^{x'+x_1+\ldots+x_{n-1}} u - \pi^*z\right)}{\theta\left(\int_x^y v - z\right)\theta\left(\int_D^{x+x_1+\ldots+x_{n-1}} v + z\right)} = c(y',x_1,\ldots,x_{n-1}) \tag{99}$$

and $c(x_1,\ldots,x_n) = 0$ if and only if $x_i = x_j'$ for some $i,j \in 1,\ldots,n$, $i \neq j$. For $m > 0$, $\hat{\theta}$ has multiplicity m on the subvariety

$$\pi^*J_0(C) + \{\int_D^{x_1+\ldots+x_n} u\} \subset J_0(\hat{C}) \text{ for any } x_1,\ldots,x_n \in \hat{C} \text{ such that}$$

$x_i = x_j'$ for m disjoint pairs (i,j) with $i \neq j$ and $i,j \in 1,\ldots,n$.

<u>Proof</u>. (97) gives (99) by replacing x_n by y' and z by $z - \int_x^y v$. For generic $z = \mathcal{A} - x - \Delta \in J(C)$ with $\mathcal{A} = \mathrm{div}_C\theta(\int_x^y v - z) \neq C$,

$$\pi^*z+x'+x_1+\ldots+x_{n-1}-D = \pi^*\mathcal{A} + x_1+\ldots+x_{n-1}-x-\hat{\Delta} \in J_0(\hat{C})$$

so that the left hand side of (99) has zeroes at $y = x_1,\ldots,x_{n-1} \in \hat{C}$ for generic $x_1,\ldots,x_{n-1} \in \hat{C}$. Since the locus $c(x_1,\ldots,x_n) = 0$ is a subvariety of $\hat{C}\times\ldots\times\hat{C}$ of codimension 1 and since we have just seen that $c(x_1,\ldots,x_{n-1},y)$ has simple zeroes at $y = x_1',\ldots,x_{n-1}'$ for generic $x_1,\ldots,x_{n-1}$, the symmetry of c implies that the variety $c = 0$

is the union of the divisors $x_i = x_j'$, $i \neq j$, in $\hat{C} \times \ldots \times \hat{C}$. The assertion concerning $\hat{\Theta}$ now follows from (97).

The section c in (97)-(99) depends of course on the choice of divisor in the class $D \in J_n(\hat{C})$; the divisor class D itself is given by (101) and (103) of the following

<u>Proposition 5.3.</u> For any partition $\{i_1,\ldots,i_{n-2m}\} \cup \{j_1,\ldots,j_{n+2m}\}$ of $\{1,2,\ldots,2n\}$ with m a non-negative integer, there is a unique half-period of the form $\begin{Bmatrix} 0 & \mu & 0 \\ 0 & \nu & 0 \end{Bmatrix}_{\hat{\tau}} \in J_0(\hat{C})$ such that

$$D - (Q_{i_1} + \ldots + Q_{i_{n-2m}}) \in \begin{Bmatrix} 0 & \mu & 0 \\ 0 & \nu & 0 \end{Bmatrix}_{\hat{\tau}} + \pi^* J_m(C) \subset J_{2m}(\hat{C}).$$

For a partition with $m > 0$, $\hat{\Theta}$ has multiplicity m on the subvariety $\begin{Bmatrix} 0 & \mu & 0 \\ 0 & \nu & 0 \end{Bmatrix}_{\hat{\tau}} + \pi^* J_0(C)$* so that, by Riemann's Theorem 1.1, $\hat{\Theta}\begin{bmatrix} 0 & \mu & 0 \\ 0 & \nu & 0 \end{bmatrix}(\int_x^y u + \pi^* e) \equiv 0$ for all $e \in \mathbb{C}^g$ and $x,y \in \hat{C}$ whenever $\begin{Bmatrix} 0 & \mu & 0 \\ 0 & \nu & 0 \end{Bmatrix}_{\hat{\tau}}$ corresponds to a "singular partition" with $m > 1$. If $\begin{Bmatrix} 0 & \mu & 0 \\ 0 & \nu & 0 \end{Bmatrix}_{\hat{\tau}}$ is a non-singular odd half-period coming from a partition with $m = 1$,

$$(100) \quad \hat{\Theta}(\int_x^y u + \pi^* e - \begin{Bmatrix} 0 & \mu & 0 \\ 0 & \nu & 0 \end{Bmatrix}_{\hat{\tau}}) = c(y',x,Q_{i_1},\ldots,Q_{i_{n-2}})\Theta(\int_d^y v + e)\Theta(\int_d^x v - e)$$

$\forall\, e \in \mathbb{C}^g$, where $d = a + \frac{1}{4}\int_{Q_{i_1}+\ldots+Q_{i_{n-2}}+4a}^{Q_{j_1}+\ldots+Q_{j_{n+2}}} v$ for any $a \in C$. Finally, if $\begin{Bmatrix} 0 & \mu & 0 \\ 0 & \nu & 0 \end{Bmatrix}_{\hat{\tau}}$ is a non-singular even half-period corresponding to a partition with $m = 0$,

$$(101) \quad D = \hat{\Delta} - \pi^*\Delta = Q_{i_1} + \ldots + Q_{i_n} + \pi^*(\tfrac{1}{4}\int_{Q_{i_1}+\ldots+Q_{i_n}}^{Q_{j_1}+\ldots+Q_{j_n}} v) + \begin{Bmatrix} 0 & \mu & 0 \\ 0 & \nu & 0 \end{Bmatrix}_{\hat{\tau}} \in J_n(\hat{C}),$$

and for all $e \in \mathbb{C}^g$ and $2\alpha, 2\beta \in \mathbb{Z}^g$,

---

* This has been proved also in [1, II, p. 22].

(102) $$\frac{\hat{\theta}\begin{bmatrix}\alpha & \mu & -\alpha\\ \beta & \nu & -\beta\end{bmatrix}(\pi^* e)}{\theta\begin{bmatrix}\alpha\\ \beta\end{bmatrix}(e+\xi)\,\theta\begin{bmatrix}\alpha\\ \beta\end{bmatrix}(e-\xi)} = c\begin{bmatrix}\mu\\ \nu\end{bmatrix} = c(Q_{i_1},\dots,Q_{i_n})\exp \tfrac{1}{4}\mu \underline{\Pi}\mu^t$$

by Prop. 5.1, where $\xi = \frac{1}{4}\int_{Q_{i_1}+\dots+Q_{i_n}}^{Q_{j_1}+\dots+Q_{j_n}} v$ and $\underline{\Pi} = (\Pi_{ij})$, for $g+1 \leq i,j \leq g+n-1$.

Proof. For notational convenience, let $Q_I = Q_{i_1} + \dots + Q_{i_{n-2m}}$ and $Q_J = Q_{j_1} + \dots + Q_{j_{n+2m}}$; for any positive divisor $X = \sum_1^m x_i$ of degree m on C, set

$$\tfrac{1}{4}(Q_J - Q_I) = X + \tfrac{1}{4}\int_{Q_I+4X}^{Q_J} v \in J_m(C)$$

where the integration is taken within C cut along its homology basis. Then the divisor class

(103) $$A = \pi^*(\tfrac{1}{4}\pi_*(Q_I - Q_J)) - Q_I + D \in J_0(\hat{C})$$

is invariant under ϕ and, from (96), satisfies

$$4A = \pi^*(Q_I - Q_J) - 4Q_I + 4D = -2Q_I - 2Q_J + 4D = 0.$$

Thus A is a quarter period of the form $\begin{Bmatrix}\alpha & \mu & -\alpha\\ \beta & \nu & -\beta\end{Bmatrix}_{\hat{\tau}} \in J_0(C)$ and, by (97),

(104) $$\frac{\hat{\theta}(\pi^* e+Y-X-A)}{\theta(e - \frac{1}{4}\int_{Q_J}^{Q_I+4X} v)\,\theta(e + \frac{1}{4}\int_{Q_J}^{Q_I+4Y} v - \pi_* A)} = c(Q_{i_1},\dots,Q_{i_{n-2m}},y_1,x_1',\dots,y_m,x_m')$$

for all $e \in \mathbb{C}^g$ and positive divisors $Y = \sum_1^m y_i$ of degree m on $\hat{C}$. Now the characteristics of $\begin{Bmatrix}\alpha & \mu & -\alpha\\ \beta & \nu & -\beta\end{Bmatrix}_{\hat{\tau}}$ must remain constant for a family of surfaces $\hat{C}_t$ obtained by pinching C along a loop homotop to

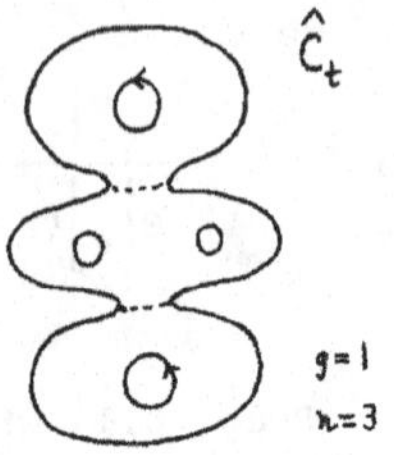

zero, enclosing and not separating the points Q_i: applying the formulas (47)-(48) of §3 to (104) with the divisors X and Y near some of the points Q_i, we have, for all $e \in \mathbb{C}^g$,

$$\frac{\Theta_\tau(e - \begin{Bmatrix}\alpha\\ \beta\end{Bmatrix})\Theta_\tau(e - \begin{Bmatrix}\alpha\\ \beta\end{Bmatrix})\Theta_s(\int_X^Y w - \begin{Bmatrix}\mu\\ \nu\end{Bmatrix})}{\Theta_\tau(e)\Theta_\tau(e - \begin{Bmatrix}2\alpha\\ 2\beta\end{Bmatrix})} = \lim_{t\to 0} c_t(Q_{i_1},\ldots,Q_{i_{n-2m}},y_1,x_1',\ldots,y_m,x_m')$$

where τ and s are the period matrices for C and the hyperelliptic Riemann surface of genus n-1 with Weierstrass points $Q_1,\ldots,Q_{2n}$. Thus $\begin{Bmatrix}\alpha\\ \beta\end{Bmatrix}_\tau = 0$ in $J_0(C)$ and $\begin{bmatrix}\mu\\ \nu\end{bmatrix}$ is the half-integer hyperelliptic characteristic of genus n-1 corresponding to the partition $\{i_1,\ldots,i_{n-2m}\} \sqcup \{j_1,\ldots,j_{n+2m}\}$ of $\{1,2,\ldots,2n\}$ according to the rule on p. 13. Formulas (100) and (102) now come from (101), (103) and (104); and the multiplicity of $\hat{\Theta}$ on the subvariety $\begin{Bmatrix}0 & \mu & 0\\ 0 & \nu & 0\end{Bmatrix}_{\hat{\tau}} + \pi^* J_0$ is computed from Cor. 5.2 or directly from (104), letting $y_k \to x_k$ for $k = 1,\ldots,m$.

In contrast to the unramified case in §4, the prime-form E(x,y) on C does not become a multiplicative $-\frac{1}{2}$ order differential on $\hat{C}$ when lifted to a multiplicative section, in x and y, of the induced bundle $\pi^* L_0^{-1}$ on $\hat{C}$. However, the pullback $[(d\pi)^* \times (d\pi)^*]E^2(x,y)$ *is* a multiplicative inverse differential on $\hat{C} \times \hat{C}$ with 2n simple poles at $x,y = Q_1,\ldots,Q_{2n}$ and double zeroes at $y = x$ and $y = x'$. We will retain the notation $E^2(x,y)$ for this multiplicative inverse differential of $x,y \in \hat{C}$, so that $\Theta^2(y-x)/E^2(x,y)$ is a bilinear differential on $C \times C$ lifted to $\hat{C} \times \hat{C}$, with double poles at $y = x,x'$ and simple zeroes at $Q_1,\ldots,Q_{2n}$ - analogous to the bilinear differential $\frac{dz(x)dz(y)}{(z(x)-z(y))^2}$ in the hyperelliptic case $\hat{C} \xrightarrow{z} C = \mathbb{P}_1(\mathbb{C})$.

<u>Proposition 5.4.</u>* Let $\begin{Bmatrix} 0 & \mu & 0 \\ 0 & \nu & 0 \end{Bmatrix}_T \in J_0(\hat{C})$ be a non-singular even half-period corresponding to a partition $\{i_1,\ldots,i_n\} \sqcup \{j_1,\ldots,j_n\}$ of $\{1,\ldots,2n\}$ as in Proposition 5.3. Then $\forall\ e \in \mathbb{C}^g$ and $x,y \in \hat{C}$,

$$(105)\quad 2\frac{\hat{\Theta}\begin{bmatrix} 0 & \mu & 0 \\ 0 & \nu & 0 \end{bmatrix}(\int_x^y u - \pi^* e)}{\hat{\Theta}\begin{bmatrix} 0 & \mu & 0 \\ 0 & \nu & 0 \end{bmatrix}(\pi^* e)\hat{E}(x,y)} = \frac{\Theta(\int_x^y v - e + \xi)}{\Theta(e-\xi)E(x,y)}\sqrt{\frac{\sigma(y)}{\sigma(x)}} + \frac{\Theta(\int_x^y v - e - \xi)}{\Theta(e+\xi)E(x,y)}\sqrt{\frac{\sigma(x)}{\sigma(y)}}$$

where $\xi = \frac{1}{4}\int_{Q_{i_1}+\ldots+Q_{i_n}}^{Q_{j_1}+\ldots+Q_{j_n}} v$, and $\sigma(x) = \sqrt{\prod_1^n \frac{E(x,Q_{j_k})}{E(x,Q_{i_k})}}$ is a section of $\pi^*(2\xi) \in J_0(\hat{C})$ with simple zeroes at the Q_{j_k} and simple poles at the Q_{i_k}. (The sign of the square root in (105) is chosen to be positive when $y = x$.)

<u>Proof.</u> First of all, the right hand side of (105) makes sense since $E^2(x,y)\sigma(x)\sigma(y)$ has double zeroes at $y = x,x'$ and double poles at the Q_{i_k}, so that $\frac{1}{E(x,y)}\sqrt{\frac{\sigma(y)}{\sigma(x)}}$ (respectively $\frac{1}{E(x,y)}\sqrt{\frac{\sigma(x)}{\sigma(y)}}$) is, for fixed $x \in \hat{C}$, a well-defined section of the bundle on $\hat{C}$ with divisor $\sum_1^n Q_{j_k} - x - x'$ (respectively $\sum_1^n Q_{i_k} - x - x'$). For fixed x,

$$\psi_1(x,y) = \frac{\Theta(\int_x^y v - e + \xi)}{\Theta(e-\xi)E(x,y)}\sqrt{\frac{\sigma(y)}{\sigma(x)}} \quad\text{and}\quad \psi_2(x,y) = \frac{\Theta(\int_x^y v - e - \xi)}{\Theta(e-\xi)E(x,y)}\sqrt{\frac{\sigma(x)}{\sigma(y)}}$$

are multiplicative half-order differentials of $y \in \hat{C}$, holomorphic except for simple poles at $y = x$ with residues 1, and simple poles at $y = x'$ with residues of opposite sign since $\sigma(x') = -\sigma(x)$. By definition of ξ, $\frac{1}{4}(\psi_1 + \psi_2)^2$ is therefore a meromorphic section of $\pi^*(2e) \otimes K_{\hat{C}}$ with all double zeroes and a double pole of residue 1

---

* For e a half-period, this appears in [29 II, p. 746]; in the hyperelliptic case see (17) in §1.

at $y = x$; as such, it is given by $\dfrac{\hat{\Theta}[\beta]^2(\int_x^y u - \pi^* e)}{\hat{\Theta}[\beta]^2(0)\hat{E}^2(x,y)}$, where the half period $\beta \in J_0(\hat{C})$ satisfies

$$\pi^* e + \hat{\Delta} + \beta = \operatorname{div}_{\hat{C}} \Theta(\int_x^y v - e + \xi) + \operatorname{div}_{\hat{C}} \frac{1}{E(x,y)} \sqrt{\frac{\sigma(y)}{\sigma(x)}} \in J_{\hat{g}-1}(\hat{C})$$

so that, by (101),

$$\beta = -\pi^* e - \hat{\Delta} + \pi^*(e - \xi + x + \Delta) + \sum_1^n Q_{j_k} - x - x' = D - \pi^* \xi - \sum_1^n Q_{i_k} = \begin{Bmatrix} 0 & \mu & 0 \\ 0 & \nu & 0 \end{Bmatrix}_{\hat{\tau}} .$$

The Prym Variety. In analogy with §4, the Prym variety $P = \{\mathcal{A} - \mathcal{A}' \mid \mathcal{A} \in J(\hat{C})\}$ is the subgroup of all points

(106) $$\begin{Bmatrix} \alpha & \mu & \alpha \\ \beta & \nu & \beta \end{Bmatrix}_{\hat{\tau}} \in J(\hat{C}) \quad \alpha, \beta \in \mathbb{R}^g \text{ and } \mu, \nu \in \mathbb{R}^{n-1} ;$$

and there is an isogeny $i\colon J(C) \times P \to J(\hat{C})$ of degree 4^g defined by $i(\mathcal{A}, \mathcal{B}) = \pi^* \mathcal{A} + \mathcal{B}$, $\mathcal{A} \in J(C)$ and $\mathcal{B} \in P$, with kernel(i) isomorphic to the group $\pi^* J \cap P$ of 4^g half-periods in $J(C)$ lifted to $J(\hat{C})$. Let us define the projections $\sigma_1\colon J(\hat{C}) \to J(C)$ by $\sigma_1(\mathcal{A}) = \pi_*(\mathcal{A})$ and $\sigma_2\colon J(\hat{C}) \to P$ by $\sigma_2(\mathcal{A}) = \mathcal{A} - \mathcal{A}'$, so that $\pi^*\sigma_1 + \sigma_2 = 2\mathrm{Id}_{J(\hat{C})}$. Then from (94) and (106), it is readily seen that $\ker \sigma_1 = P$ and $\ker \sigma_2 = \bigsqcup_{\gamma \in \Gamma} (\gamma + \pi^* J(C))$, where Γ is the group of 4^{n-1} half-periods of the form $\begin{Bmatrix} 0 & \mu & 0 \\ 0 & \nu & 0 \end{Bmatrix}_{\tau} \in J(\hat{C})$ with 2μ and $2\nu \in \mathbb{Z}^{n-1}$; and this implies that $J(\hat{C})/\pi^* J(C)$ is, under σ_2, isogenous to P with kernel Γ. The projections σ_1 and σ_2 will be lifted to $\mathbb{C}^{\hat{g}}$ by setting, for $\hat{z} \in \mathbb{C}^{\hat{g}}$,

$$z = \sigma_1(\hat{z}) = (z_1, \ldots, z_g) \in \mathbb{C}^g, \qquad z_\alpha = \hat{z}_\alpha - \hat{z}_{\alpha'}$$

and

$$\hat{s} = \sigma_2(\hat{z}) = (s_1, \ldots, s_g, 2s_{g+1}, \ldots, 2s_{g+n-1}, s_1, \ldots, s_g) \in \mathbb{C}^{\hat{g}}$$

where, by (94) and (106),

$$s_\alpha = \hat{z}_\alpha + \hat{z}_{\alpha'}, \quad 1 \le \alpha \le g \quad \text{and} \quad 2s_i = 2\hat{z}_i, \quad g+1 \le i \le g+n-1$$

are the coordinates of a point in the universal cover $\tilde{P} \subset \mathbb{C}^{\hat{g}}$ of $P \subset J(\hat{C})$. Thus

(107) $$2\hat{z} = \pi^*\sigma_1(\hat{z}) + \sigma_2(\hat{z}) = \pi^*z + \hat{s} = \pi^*z + \phi(s) \in \mathbb{C}^{\hat{g}}$$

where ϕ is the isomorphism from $\mathbb{C}^{g+n-1}$ onto $\tilde{P} \subset \mathbb{C}^{\hat{g}}$ defined by sending $s = (s_1,\ldots,s_{g+n-1}) \in \mathbb{C}^{g+n-1}$ to

(108) $$\phi(s_1,\ldots,s_g,\ldots,s_{g+n-1}) = (s_1,\ldots,s_g,2s_{g+1},\ldots,2s_{g+n-1},s_1,\ldots,s_g) \in \mathbb{C}^{\hat{g}}.$$

If the isomorphism $\psi\colon \mathbb{C}^{g+n-1} \xrightarrow{\sim} \tilde{P}$ is given by

$$\psi(s_1,\ldots,s_g,\ldots,s_{g+n-1}) = (s_1,\ldots,s_g,s_{g+1},\ldots,s_{g+n-1},s_1,\ldots,s_g) \in \mathbb{C}^{\hat{g}}$$

then the Riemann matrix $\hat{\tau}$ restricted to $\tilde{P}$ and pulled back by ψ becomes on $\mathbb{C}^{g+n-1}$ twice the symmetric matrix Π given by (92) which has negative definite real part since $\hat{\tau}$ does; also $\phi\psi^t(\hat{s}) = 2\hat{s}$ for any $\hat{s} \in \tilde{P}$, so that $\forall\ \alpha,\beta \in \mathbb{R}^g$ and $\mu,\nu \in \mathbb{R}^{n-1}$, (108) gives

$$\phi\begin{Bmatrix}\alpha & \mu\\ \beta & \nu\end{Bmatrix}_{\Pi} = \begin{Bmatrix}\alpha & \mu & \alpha\\ \beta & 2\nu & \beta\end{Bmatrix}_{\hat{\tau}} \in \tilde{P}.$$

Consequently, if $P_0 = \mathbb{C}^{g+n-1}/(2\pi i I,\Pi)$ is the $g+n-1$ dimensional principally polarized Abelian variety formed from the Riemann matrix Π, ϕ induces an isogeny $\phi\colon P_0 \to P$ of degree 2^{n-1} with kernel given by the group of half-periods of the form $\begin{Bmatrix}0 & 0\\ 0 & \kappa\end{Bmatrix}_{\Pi}$ with $2\kappa \in (\mathbb{Z}/2\mathbb{Z})^{n-1}$. Proceeding exactly as in Prop. 4.8, one has:

<u>Proposition 5.5.</u> Let η be the Riemann theta-function for P_0 and for $\hat{z} \in \mathbb{C}^{\hat{g}}$, set $z = \pi_*(\hat{z}) \in \mathbb{C}^g$ and $s = \phi^{-1}(\sigma_2(\hat{z})) \in \mathbb{C}^{g+n-1}$ as in (107). Then for all $a,b,c,d \in \mathbb{R}^g$ and $\mu,\nu \in \mathbb{R}^{n-1}$,

(109) $$\hat{\theta}_{\hat{\tau}}\begin{bmatrix}a & \mu & c\\ b & \nu & d\end{bmatrix}(\hat{z}) = \sum_{2\delta\in(\mathbb{Z}/2\mathbb{Z})^g} \theta_{2\tau}\begin{bmatrix}\frac{a-c}{2}+\delta\\ b-d\end{bmatrix}(z)\,\eta_{2\pi}\begin{bmatrix}\frac{a+c}{2}+\delta & \mu\\ b+d & \nu\end{bmatrix}(s),$$

which inverts to:

$$(110)\quad 2^g\Theta_{2\tau}\begin{bmatrix}a\\b\end{bmatrix}(z)\eta_{2\pi}\begin{bmatrix}c & \mu\\ d & \nu\end{bmatrix}(s) = \sum_{2\varepsilon\in(\mathbb{Z}/2\mathbb{Z})^g} e^{-4\pi i\varepsilon\cdot c}\,\hat{\Theta}_{\hat{\tau}}\begin{bmatrix}c+a & \mu & c-a\\ \frac{d+b}{2}+\varepsilon & \nu & \frac{d-b}{2}+\varepsilon\end{bmatrix}(\hat{z}).$$

In particular, $\hat{\Theta}^2_{\hat{\tau}}(\hat{z})/\Theta_\tau(\sigma_1(\hat{z}))\eta_\pi(\phi^{-1}\sigma_2(\hat{z}))$ is a well-defined meromorphic function on $J_0(\hat{C})$.

Analogous to Cor. 4.9 is:

<u>Corollary 5.6</u>. If $e \in \mathbb{C}^{g+n-1}$ and $a \in \hat{C}$, then either $\eta(\int_a^x w - e) \equiv 0$ for all $x \in \hat{C}$ or $\mathrm{div}_{\hat{C}}\eta(\int_a^x w - e) = \zeta$ is of degree $2(g+n-1)$ satisfying

$$(111)\qquad \phi(e) = \zeta + a' - a - (Q_1+\ldots+Q_{2n}) - \pi^*\Delta \in J_0(\hat{C}),$$

and $\pi_*\zeta$ is the divisor of zeroes of a differential of the third kind on C with at most simple poles at $Q_1,\ldots,Q_{2n}$.

<u>Proof</u>. Let $\mathrm{div}_C\Theta(\sigma_1(x-a)) = \mathcal{B}$ and $\mathrm{div}_{\hat{C}}\hat{\Theta}(x-a-\hat{e}) = \mathcal{A}$ where $\hat{e} = \frac{1}{2}\phi(e) \in \mathbb{C}^{\hat{g}}$; then (10) implies

$$0 = \mathcal{B} - \pi_*a - \Delta \in J_0(C) \quad\text{and}\quad \hat{e} = \mathcal{A} - a - \hat{\Delta} \in J_0(\hat{C}).$$

But $\zeta + \pi^*\mathcal{B} = 2\mathcal{A} \in J_{2\hat{g}}(\hat{C})$ by Prop. 5.5, so (96) gives

$$2\hat{e} = \zeta + \pi^*(\pi_*a + \Delta) - 2a - 2\hat{\Delta} = \zeta + a' - a - (Q_1+\ldots+Q_{2n}) - \pi^*\Delta.$$

Since $\sigma_1(\hat{e}) = 0$, $\pi_*\zeta - (Q_1+\ldots+Q_{2n}) = \pi_*\pi^*\Delta = 2\Delta = K_C$ which gives the last assertion.

When there are only two branch points, the differentials of the third kind arising in Cor. 5.6 are given explicitly on p. 101.

<u>Proposition 5.7</u>. For $x_1,\ldots,x_n \in \hat{C}$, $2\rho \in \mathbb{Z}^g$ and c the section of Prop. 5.1,

$$(112)\qquad \eta_{2\pi}\begin{bmatrix}\rho & 0\\ 0 & 0\end{bmatrix}\left(\int_0^{x_1+\ldots+x_n} w\right) = c(x_1,\ldots,x_n)\Theta_{2\tau}\begin{bmatrix}\rho\\0\end{bmatrix}\left(\int_0^{x_1+\ldots+x_n} v\right).$$

If the half-period $\left\{\begin{smallmatrix} 0 & \mu & 0 \\ 0 & \nu & 0 \end{smallmatrix}\right\}_{\hat{\tau}}$ corresponds to the partition

$\{i_1, \ldots, i_{n-2m}\} \cup \{j_1, \ldots, j_{n+2m}\}$ as in Prop. 5.3, then

$$\eta_{2\pi}\begin{bmatrix} \rho & \mu \\ 0 & \nu \end{bmatrix}(\int_x^y w) \equiv 0 \quad \text{on} \quad \hat{C} \times \hat{C} \quad \text{if} \quad m > 1,$$

$$\eta_{2\pi}\begin{bmatrix} \rho & 0 \\ 0 & 0 \end{bmatrix}(\int_x^y w - \left\{\begin{smallmatrix} 0 & \mu \\ 0 & \nu \end{smallmatrix}\right\}) = c(y', x, Q_{i_1}, \ldots, Q_{i_{n-2}}) \theta_{2\tau}\begin{bmatrix} \rho \\ 0 \end{bmatrix}(\int_{2d}^{x+y} v)$$

for $m = 1$ and d as in (100); and

$$(113) \quad \frac{\eta_{2\pi}\begin{bmatrix} \rho & \mu \\ 0 & \nu \end{bmatrix}(\int_x^y w)}{\hat{E}(x,y)} = \frac{c\begin{bmatrix} \mu \\ \nu \end{bmatrix}}{2\,E(x,y)}\left(\theta_{2\tau}\begin{bmatrix} \rho \\ 0 \end{bmatrix}(\int_x^y v + 2\xi)\sqrt{\frac{\sigma(y)}{\sigma(x)}} + \theta_{2\tau}\begin{bmatrix} \rho \\ 0 \end{bmatrix}(\int_x^y v - 2\xi)\sqrt{\frac{\sigma(x)}{\sigma(y)}}\right)$$

for $m = 0$ and $c\begin{bmatrix} \mu \\ \nu \end{bmatrix}$, ξ given by (102), so that

$$\eta_{2\pi}\begin{bmatrix} \rho & \mu \\ 0 & \nu \end{bmatrix}(0) = c\begin{bmatrix} \mu \\ 0 \end{bmatrix}\theta_{2\tau}\begin{bmatrix} \rho \\ 0 \end{bmatrix}(\frac{1}{2}\int_{Q_{i_1} + \ldots + Q_{i_n}}^{Q_{j_1} + \ldots + Q_{j_n}} v).$$

Proof. For any $e \in \mathfrak{C}^g$ and $2\rho \in \mathbf{Z}^g$, (110), (98) and (3) together imply that

$$2^g\,\theta_{2\tau}\begin{bmatrix} \rho \\ 0 \end{bmatrix}(2e)\,\eta_{2\pi}\begin{bmatrix} \rho & 0 \\ 0 & 0 \end{bmatrix}(\int_D^{x_1 + \ldots x_n} w) = \sum_{2\epsilon\,\in\,(\mathbf{Z}/2\mathbf{Z})^g}(-1)^{4\rho\cdot\epsilon}\,\hat{\theta}_{\hat{\tau}}\begin{bmatrix} 0 & 0 & 0 \\ \epsilon & 0 & \epsilon \end{bmatrix}(\pi^*e + \phi(\frac{1}{2}\int_D^{x_1 + \ldots x_n} w))$$

$$= c(x_1, \ldots, x_n)\sum_{2\epsilon\,\in\,(\mathbf{Z}/2\mathbf{Z})^g}(-1)^{4\rho\cdot\epsilon}\,\theta_\tau\begin{bmatrix} 0 \\ \epsilon \end{bmatrix}(e - \frac{1}{2}\int_D^{x_1 + \ldots + x_n} v)\,\theta_\tau\begin{bmatrix} 0 \\ \epsilon \end{bmatrix}(e + \frac{1}{2}\int_D^{x_1 + \ldots + x_n} v)$$

$$= 2^g c(x_1, \ldots, x_n)\,\theta_{2\tau}\begin{bmatrix} \rho \\ 0 \end{bmatrix}(2e)\,\theta_{2\tau}\begin{bmatrix} \rho \\ 0 \end{bmatrix}(\int_D^{x_1 + \ldots + x_n} v)$$

which gives (112). For $\begin{bmatrix} 0 & \mu & 0 \\ 0 & \nu & 0 \end{bmatrix}$ a non-singular odd or even half-period,

repeat the above argument with (100) or (105) and (102) instead of

(98); and when $\begin{bmatrix} 0 & \mu & 0 \\ 0 & \nu & 0 \end{bmatrix}$ is a singular half-period, use (110) together

with the fact that $\hat{\theta}\begin{bmatrix} 0 & \mu & 0 \\ 0 & \nu & 0 \end{bmatrix}(\int_x^y u - \pi^*e) \equiv 0$ on $\hat{C} \times \hat{C}$ for all $e \in \mathfrak{C}^g$

by Prop. 5.3.

From (113), (109) and (4) we have

Corollary 5.8. For $s \in \mathbb{C}^{g+n-1}$ and $\forall\, x,y \in \hat{C}$,

$$\eta\left(\int_x^y w - s\right)\eta(s) = \frac{\hat{E}(x,y)}{2E(x,y)} \sum_\mu c\begin{bmatrix}\mu\\0\end{bmatrix} \left\{ \hat{\Theta}\begin{bmatrix}0 & \mu & 0\\ 0 & 0 & 0\end{bmatrix}\left(\int_x^y u - \phi(s) + \pi^*\xi\right)\sqrt{\frac{\sigma(y)}{\sigma(x)}} + \right.$$

$$(114) \qquad \left. \hat{\Theta}\begin{bmatrix}0 & \mu & 0\\ 0 & 0 & 0\end{bmatrix}\left(\int_x^y u - \phi(s) - \pi^*\xi\right)\sqrt{\frac{\sigma(x)}{\sigma(y)}} \right\}$$

where the summation is extended over all non-singular even half-periods of the form $\left\{\begin{matrix}0 & \mu & 0\\ 0 & 0 & 0\end{matrix}\right\}_{\hat{\tau}}$ with corresponding ξ,σ as in Prop. 5.4. When $y = x$, (114) and (109) give

$$\eta^2(s) = \sum_\mu c\begin{bmatrix}\mu\\0\end{bmatrix}\hat{\Theta}\begin{bmatrix}0 & \mu & 0\\ 0 & 0 & 0\end{bmatrix}(\phi(s) + \pi^*\xi) \qquad \forall\, s \in \mathbb{C}^{g+n-1}.$$

Now for any half-period $\left\{\begin{matrix}\alpha\\ \beta\end{matrix}\right\}_\tau \in \mathbb{C}^g$, $\phi^{-1}\pi^*\left\{\begin{matrix}\alpha\\ \beta\end{matrix}\right\}_\tau \in \mathbb{C}^{g+n-1}$ is a half Π-period which is even or odd with $\left\{\begin{matrix}\alpha\\ \beta\end{matrix}\right\}_\tau$: by setting $s - \phi^{-1}\left(\pi^*\left\{\begin{matrix}\alpha\\ \beta\end{matrix}\right\}_\tau\right) = \frac{1}{2}\int_x^y w$, 0 and $\frac{1}{2}\int_x^z w$ (letting $z \to x$), the above corollary, together with (102) and (105), gives relations analogous to the Schottky-relations (83), (80) and (85) respectively in the unramified case; for the form of the relations in the simplest case of 2 branch points, see (117)-(120) below. In addition, there is the following special relation for the odd $\Theta - \eta$ functions, due to Schottky-Jung [29 I, p. 292].

Proposition 5.9. For any non-singular odd half-period $\alpha = \left\{\begin{matrix}\delta\\ \varepsilon\end{matrix}\right\}_\tau \in J_0(C)$, let $\underset{\sim}{\alpha} = \left\{\begin{matrix}\delta & 0\\ \varepsilon & 0\end{matrix}\right\}_\Pi \in P_0$ be the corresponding odd half Prym period with $\phi(\underset{\sim}{\alpha}) = \pi^*\alpha$. Then $\forall\, x,y \in \hat{C}$,

$$(115) \qquad 2\,\frac{E^2(x,y)\,\eta[\underset{\sim}{\alpha}]\left(\int_x^y w\right)}{\hat{E}^2(x,y)\,\Theta[\alpha]\left(\int_x^y v\right)} = \frac{G[\underset{\sim}{\alpha}](x)}{H[\alpha](x)} + \frac{G[\underset{\sim}{\alpha}](y)}{H[\alpha](y)}$$

where $H[\alpha] = \sum_1^g \frac{\partial\theta[\alpha]}{\partial z_j}(0)v_j$ and $G[\underline{\alpha}] = \sum_1^{g+n-1} \frac{\partial\eta[\underline{\alpha}]}{\partial s_j}(0)w_j$.

<u>Proof</u>. By Prop. 5.5, $\hat{E}^{-2}(x,y)\eta[\underline{\alpha}](\int_x^y w)\theta[\alpha](\int_x^y v)$ is a symmetric bilinear holomorphic differential on $\hat{C} \times \hat{C}$ which can be written as

$$\Omega_1(x,y) + \Omega_2(x,y)\phi(x) + \Omega_3(x,y)\phi(y) + \Omega_4(x,y)\phi(x)\phi(y)$$

where ϕ is defined on p. 86, and the $\Omega_k(x,y)$ are meromorphic differentials on $C \times C$. Replacing x,y by x',y', the symmetry of $\hat{E}$ and the condition that α is odd implies $\Omega_1(x,y) + \Omega_4(x,y)\phi(x)\phi(y) = 0$. By fixing x and sending y to y', it follows that $\Omega_2(x,y)\phi(x)$ is a holomorphic differential in y vanishing at the double zeroes of $H[\alpha](y)$, so that $\Omega_2(x,y) = \psi(x)H[\alpha](y)$ for some meromorphic differential $\psi(x)$ since $i(\frac{1}{2}\mathrm{div}\, H[\alpha]) = 1$. Likewise $\Omega_3(x,y) = \Omega_2(y,x) = \psi(y)H[\alpha](x)$ so

$$\eta[\underline{\alpha}](\int_x^y w)\theta[\alpha](\int_x^y v) = \hat{E}^2(x,y)\Big(H[\alpha](x)\phi(y)\psi(y) + H[\alpha](y)\phi(x)\psi(x)\Big).$$

Letting $y \to x$, $G[\underline{\alpha}](x)H[\alpha](x) = 2H[\alpha](x)\phi(x)\psi(x)$ and therefore

$$2\eta[\underline{\alpha}](\int_x^y w)\theta[\alpha](\int_x^y v) = \hat{E}^2(x,y)\Big(H[\alpha](x)G[\underline{\alpha}](y) + H[\alpha](y)G[\underline{\alpha}](x)\Big)$$

which, by (19), implies (115).

<u>Ramified Coverings with Two Branch Points</u>. When the double covering $\hat{C} \to C$ has only two branch points a and b, the g-dimensional Prym variety $P_0 \overset{\phi}{\simeq} P$ is principally polarized and the relations between the $\hat{\theta}$, θ and η functions have a particularly simple form, due to the fact that the η-divisor on $\hat{C}$ is a translate of the $\hat{\theta}$-divisor on $\hat{C}$. By (101) of Prop. 5.3, the divisor class D in this case is given by

$$D = \hat{\Delta} - \pi^*\Delta = a + \pi^*(\tfrac{1}{4}\int_a^b v) = b - \pi^*(\tfrac{1}{4}\int_a^b v) = a + \pi^*\xi = D'$$

so that $\forall\, x \in \hat{C}$,

$$\int_D^x w = \int_a^x w = \int_b^x w = \tfrac{1}{2}\int_{x'}^x w \quad \text{and} \quad \int_D^x v = \int_a^x v - \tfrac{1}{2}\int_a^b v = \int_b^x v + \tfrac{1}{2}\int_a^b v .$$

The section $\sigma(x)$ of Prop. 5.4 for the partition $\{a\} \cup \{b\}$ is $\sigma(x) = \sqrt{\frac{E(x,b)}{E(x,a)}}$, while the section $c(x) = c(x')$ of Prop. 5.1 has no zeroes or poles on $\hat{C}$.

<u>Proposition 5.10</u>. For all $s \in \mathbb{C}^g$, $x,y \in \hat{C}$, and half-integer characteristics γ,

$$(116)\qquad \frac{\eta\left(\frac{1}{2}\int_D^y w + s\right)\eta\left(\frac{1}{2}\int_D^y w - s\right)}{\hat{\theta}\left(\pi^*\left(\frac{1}{2}\int_D^y v\right) - \phi(s)\right)} = c(y) = \frac{\eta\left(\int_x^y w - s\right)\eta\left(\int_D^x w + s\right)}{\hat{\theta}\left(\int_{x+0}^{y+x'} u - \phi(s)\right)}$$

$$(117)\qquad \frac{\eta[\gamma]\left(\frac{1}{2}\int_{a+b}^{x+y} w\right)\eta[\gamma]\left(\frac{1}{2}\int_x^y w\right)}{\theta[\gamma]\left(\frac{1}{2}\int_{a+b}^{x+y} v\right)\theta[\gamma]\left(\frac{1}{2}\int_x^y v\right)} = c(x)c(y) \quad \text{and} \quad \frac{\eta[\gamma]^2\left(\frac{1}{2}\int_D^x w\right)}{\theta[\gamma]\left(\frac{1}{2}\int_a^x v\right)\theta[\gamma]\left(\frac{1}{2}\int_b^x v\right)} = c(x)c(a)$$

<u>Proof</u>. To establish the left hand side of (116), replace s by $s + \tfrac{1}{2}\int_D^y w$ in Cor. 5.8 and let $x \to a$, making use of the fact that by (105) and (97):

$$(118)\qquad \lim_{x\to a} \tfrac{1}{2}\, \frac{\hat{E}(x,y)}{E(x,y)}\sqrt{\frac{\sigma(x)}{\sigma(y)}} = \frac{\hat{\theta}\left(\int_a^y u - \pi^* e\right)}{\hat{\theta}(\pi^* e)}\; \frac{\theta(e+\xi)}{\theta\left(\int_a^y v - e - \xi\right)} = \frac{c(y)}{c(a)} \qquad \forall\, y \in \hat{C}.$$

The right-hand side of (116) comes by replacing s with $s - \tfrac{1}{2}\int_{x+D}^{y+x'} w$ in the left-hand side of (116). Finally (117) is obtained by setting $s = \{\gamma\} + \tfrac{1}{2}\int_x^y w \in \mathbb{C}^g$ in (116) and using (98) of Prop. 5.1.

From (116) and (111) we see that if $x \in \hat{C}$ is fixed and $\operatorname{div}_{\hat{C}} \eta\left(\int_x^y w - s\right) = \zeta \neq \hat{C}$ for $s \in P_0$, then $i(\zeta) = 0$ if and only if

$a,b \notin \zeta$; equivalently, if $\mathrm{div}_{\hat{C}}\eta(\tfrac{1}{2}\int_{y'}^{y} w - s) = \mathcal{A} \neq \hat{C}$, then $\phi(s) = \mathcal{A} - b - a - \pi^*\Delta$ by (111) with $i(\mathcal{A}) = 0$ iff $\eta(s) \neq 0$. By Cor. 5.6, the g-dimensional linear series on C, generated by $\eta(\tfrac{1}{2}\int_{x}^{x'} w - s)\eta(\tfrac{1}{2}\int_{x}^{x'} w + s)$ for $s \in \mathbb{C}^g$, cuts out the divisors of zeroes of differentials of the third kind on C with at most simple poles at a and b, and which are holomorphic for $s \in (\eta)$: these are given explicitly by substituting $s + \tfrac{1}{2}\int_{x}^{x'} w$ for s in (114) and letting $y \to x'$:

$$\frac{4\eta(\tfrac{1}{2}\int_{x}^{x'} w + s)\eta(\tfrac{1}{2}\int_{x}^{x'} w - s)}{ic(a)\hat{E}(x',x)} = \hat{\theta}(\pi^*\xi+\phi(s))w_{b-a}(x) + \sum_{\alpha=1}^{g} \frac{\partial\hat{\theta}}{\partial z_\alpha}(\pi^*\xi+\phi(s))v_\alpha(x).$$

The differential is holomorphic iff $s \in (\eta)$ since $\hat{\theta}(\pi^*\xi + \phi(s)) = c(a)^{-1}\eta^2(s)$ by Cor. 5.8 (see also (88)).

<u>Proposition 5.11</u>. For any odd half-period α and $\forall\, x,y \in \hat{C}$,

$$G[\alpha](x)\eta[\alpha](\int_x^y w) = \frac{c^2(a)}{4}\frac{\hat{E}^2(x,y)}{E^2(x,y)}\left\{\Theta[\alpha](\int_x^y v)\frac{\Theta[\alpha]^2(\int_D^x v)E(a,b)}{\Theta[\alpha](2\xi)E(x,a)E(x,b)}\right.$$

$$\left. + H[\alpha](x)\left[\Theta[\alpha](\int_x^y v + 2\xi)\frac{\sigma(y)}{\sigma(x)} + \Theta[\alpha](\int_x^y v - 2\xi)\frac{\sigma(x)}{\sigma(y)}\right]\right\}$$

while for any even half-period β,

$$\eta[\beta](0)\eta[\beta](\int_x^y w) = \frac{c^2(a)}{4}\frac{\hat{E}^2(x,y)}{E^2(x,y)}\left\{2\Theta[\beta](2\xi)\Theta[\beta](\int_x^y v)\right.$$

$$\left. + \Theta[\beta](0)\left[\Theta[\beta](\int_x^y v + 2\xi)\frac{\sigma(y)}{\sigma(x)} + \Theta[\beta](\int_x^y v - 2\xi)\frac{\sigma(x)}{\sigma(y)}\right]\right\}.$$

In particular,

$$(119)\qquad \frac{G[\alpha](x)\eta[\alpha](\int_D^x w)}{H[\alpha](x)\Theta[\alpha](\int_D^x v)} = c^2(x), \quad \frac{G[\alpha]^2(x)}{H[\alpha](x)} = \frac{c^2(a)}{2}\frac{\Theta[\alpha]^2(\int_D^x v)E(a,b)}{\Theta[\alpha](2\xi)E(x,a)E(x,b)}$$

$$(120)\qquad \frac{\eta[\beta](0)\eta[\beta](\int_D^x w)}{\Theta[\beta](0)\Theta[\beta](\int_D^x v)} = c^2(x), \qquad \frac{\eta[\beta]^2(0)}{\Theta[\beta](0)\Theta[\beta](\frac{1}{2}\int_a^b v)} = c^2(a) = c^2(b).$$

Proof. Set $s - \frac{1}{2}\int_x^z w = \{\alpha\}$ or $\{\beta\}$ in (114) of Cor. 5.8, apply (102) and (105) and let $z \to x$, observing that $d \ln \sigma(x) = \frac{1}{2}\omega_{b-a}(x)$ by (21). The equations (119)-(120) can be proved by letting $y \to x$ and $y \to a$, making use of (118) and Cor. 2.11.

Since $G[\alpha](x)/H[\alpha](x)$ by Prop. 5.11 has at most simple poles, the index of speciality of the divisor $a+b+\pi^*(\frac{1}{2}\mathrm{div}_C H[\alpha])$ of degree $\hat{g}$ on C is positive for all odd α; these special divisors have already appeared in Prop. 5.9, which for the case of two branch points becomes

Corollary 5.12. For any odd half-period α and $\forall\, x,y \in \hat{C}$,

$$\frac{2}{\sigma_\alpha}\,\frac{E^2(x,y)}{\hat{E}^2(x,y)}\,\frac{\eta[\alpha](\int_x^y w)}{\Theta[\alpha](\int_x^y v)} = \frac{\Theta[\alpha](\int_D^x v)}{\sqrt{\Theta[\alpha](\int_a^x v)\Theta[\alpha](\int_b^x v)}} + \frac{\Theta[\alpha](\int_D^y v)}{\sqrt{\Theta[\alpha](\int_a^y v)\Theta[\alpha](\int_b^y v)}}$$

where σ_α is a constant depending on α (and $\hat{C}$) and satisfying

$$(121)\qquad \sigma_\alpha^2 = \frac{c^2(a)}{2}\,\frac{\Theta[\alpha](\int_a^b v)}{\Theta[\alpha](\frac{1}{2}\int_a^b v)} \quad\text{and}\quad \sigma_\alpha = c^2(x)\,\frac{\sqrt{\Theta[\alpha](\int_a^x v)\Theta[\alpha](\int_b^x v)}}{\eta[\alpha](\int_a^x w)}$$

for all $x \in \hat{C}$.

Proof. Since $\dfrac{G[\alpha](x)}{H[\alpha](x)}\,\dfrac{\sqrt{\Theta[\alpha](\int_a^x v)\Theta[\alpha](\int_b^x v)}}{\Theta[\alpha](\int_D^x v)}$ is, by (119), a meromorphic function on C with poles only at the zeroes of $\Theta[\alpha](\int_D^x v)$, it must be a constant σ_α independent of x which, substituted into (115), gives the first equation above; (121) comes from a Taylor expansion of (119) near $x = a$.

This corollary also follows from Prop. 5.11 and the identity

$$\frac{\theta[e](\frac{1}{2}\int_{a+b}^{2x} v)\,\theta[e](\frac{1}{2}\int_{a+b}^{2y} v)\,E(x,y)\,E(a,b)}{\theta[e](\frac{1}{2}\int_a^b v)\,E(x,a)\,E(x,b)\,E(y,a)\,E(y,b)} = \frac{\theta[e](\int_x^y v + \frac{1}{2}\int_a^b v)}{E(y,a)\,E(x,b)} - \frac{\theta[e](\int_y^x v + \frac{1}{2}\int_a^b v)}{E(y,b)\,E(x,a)}$$

valid, by (45), for all $x,y,a,b \in C$ and half-periods e. Now suppose that a and b $\neq$ a are such that $e + \frac{1}{2}\int_a^b v \in (\Theta)$ for some half-period e; then by Cor. 2.11, $\Theta[e](\frac{1}{2}\int_{a+b}^{2x} v) \equiv 0$ on C and the symmetric holomorphic differential $\Theta[e](\int_x^y v + \frac{1}{2}\int_a^b v)\Theta[e](\int_x^y v - \frac{1}{2}\int_a^b v)E(x,y)^{-2}$ on $C \times C$ has, in both x and y, g-2 double zeroes and simple zeroes at a and b. In case $e = \beta$ is an even half-period, $\eta[\beta](0) = 0$ from (120), $\{\beta\}_\pi$ is a singular point of (η), and the differential $\eta[\beta]^2(\int_D^x w)/c^4(x)E(x,a)E(x,b)$ on C is holomorphic with zeroes at a and b and g-2 double zeroes. On the other hand, when $e = \alpha$ is an odd half-period, the constant σ_α in (121) is infinite, $\eta[\alpha](\int_D^x w) \equiv 0$ on $\hat{C}$ and $G[\alpha]^2(x)/H[\alpha](x)$ is by (119) a holomorphic differential on C vanishing at a and b and with g-2 double zeroes. For fixed $a \in C$, the number of points $b \in C$ such that $\Theta[\gamma](\frac{1}{2}\int_a^b v) = 0$ for some half-period γ is $4^{g-1}g = \deg \mathrm{div}_C \prod_{2\gamma=0\in J_0} \Theta[\gamma](\frac{1}{2}\int_a^x v)$; $\frac{1}{2}(4^g - 2^g)$ of these are the simple zeroes b = a for γ an odd half-period, while the remaining b are double zeroes by, say, (83).

It is of some interest to see how the double covering $\hat{C}$ degenerates when the points a and b approach each other along some path in C; the following examples illustrate the essentially two possible types of degeneracy which can arise:

<u>Example 1</u>. If $\mathcal{C}$ is a family of ramified double coverings $\hat{C}_t$ of C with ramification points a_t and b_t approaching a point $p \in C$ so that the slit $a_t b_t$ in the limit becomes a cycle homologous to, say,

the cycle A_1 in C, then the limiting surface $\hat{C}_0$ has genus $2g-1$ and is an unramified double cover of C defined by the characteristic $\begin{bmatrix} 0 & 0 & \cdots & 0 \\ \frac{1}{2} & 0 & \cdots & 0 \end{bmatrix}$ as in §4, with an ordinary double point where the surface crosses itself at p and p', the image of p under the involution on $\hat{C}_0$. Such a family $\mathcal{C}$ will be constructed as in §3, p. 50, starting with $\hat{C}_0$ and $p,p' \in \hat{C}_0$ and pinching to a point a cycle homologous to $A_1(t) + A_{1'}(t)$ in such a way that, near p, $\mathcal{C}$ is the non-singular surface $x^2 - y^2 = t$, a_t and b_t are the points $x = -\sqrt{t}$ and $x = \sqrt{t}$, and p (resp. p') is the point $x = y = t = 0$ on the surface $x = y$ (resp. $x = -y$). Since

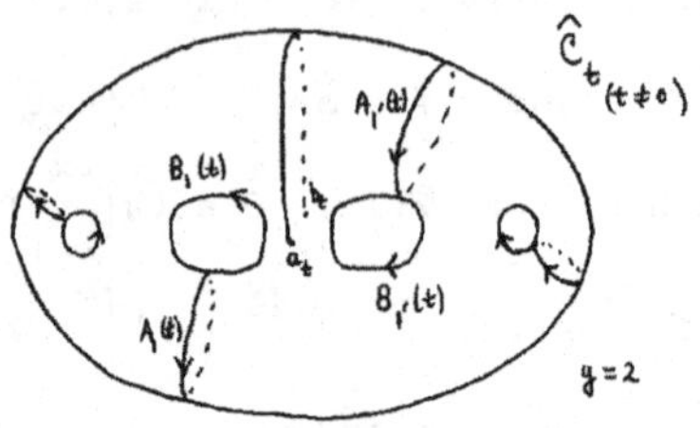

$\int_{A_1(t)+A_{1'}(t)} u_1(x,t) = \int_{A_1(t)+A_{1'}(t)} u_{1'}(x,t) = 2\pi i$ for $t \neq 0$, the normalized differentials $u_\alpha(x,t)$ with symmetry (90) on $\hat{C}_t$ have, away from the pinched region, expansions of the form

$$(122)\quad \begin{aligned} u_1(x,t) &= \omega_{p-p'}(x) + u_1(x) + O(t), & u_{1'}(x,t) &= \omega_{p-p'}(x) + O(t) \\ u_\alpha(x,t) &= u_\alpha(x) + O(t), & u_{\alpha'}(x,t) &= -u_{\alpha'}(x) + O(t), \quad 1 < \alpha \leq g \end{aligned}$$

where $u_1(x)$, $u_\alpha(x)$ and $u_{\alpha'}(x)$ are a canonical basis of $H^0(\hat{C}_0, \Omega^1_{\hat{C}_0})$ with respect to the involution as in (58), $\omega_{p-p'}(x) = -u_1(x) - \omega_{p-p'}(x')$ is the normalized differential of the third kind on $\hat{C}_0$ with simple poles of residue $+1,-1$ at p,p', and the expressions $O(t)$ are holomorphic differentials on $\hat{C}_t$ outside the pinched region which tend to zero with t. Since $\int_{B_\alpha} 2\omega_{p-p'} + \int_{B_\alpha} u_1 = \int_{p'}^{p} w_\alpha$ for the Prym differentials $w_2,\ldots,w_g$ on $\hat{C}_0$, the Prym matrix $\Pi(t)$ of $\hat{C}_t$ has, by (122), an expansion

$$\Pi(t) = \left(\begin{array}{c|c} 2\ln t + c_1 & \int_{p'}^{p} w_\beta \\ \hline \int_{p'}^{p} w_\alpha & \Pi_{\alpha\beta} \end{array}\right) + O(t) \quad , \quad 1 < \alpha,\beta \leq g$$

for some constant c_1, where $(\Pi_{\alpha\beta})$ is the Prym period matrix for $\hat{C}_0$ and $O(t)$ is a matrix satisfying $\lim_{t\to 0} O(t) = 0$. Thus $\forall\ (w_1,\ldots,w_g) \in \mathbb{C}^g$,

$$\eta_{2\pi(t)}(w_1,\ldots,w_g) = \sum_{(m_1,\ldots,m_g)\in \mathbb{Z}^g} \exp\left\{ \sum_1^g m_i\pi_{ij}(t)m_j + \sum_1^g m_i w_i \right\}$$

$$= \sum_{\substack{\bar{m}\in\mathbb{Z}^{g-1}\\ m_1\in\mathbb{Z}}} t^{2m_1^2} \exp\left\{ \bar{m}\pi\bar{m}^t + \sum_2^g m_i w_i + m_1(m_1c_1 + w_1 + 2\sum_2^g m_i\int_{p'}^{p} w_i) + O(t)\right\}$$

so that $\lim_{t\to 0} \eta_{2\pi(t)}(w_1,\ldots,w_g) = \eta_{2\pi}(w_2,\ldots,w_g)$ and, from Prop. 5.7 and (84),

$$\lim_{t\to 0} c_t(a_t) = \lim_{t\to 0} \frac{\eta_{2\pi(t)}(0)}{\Theta_{2\tau}(\tfrac{1}{2}\int_{a_t}^{b_t} v)} = \frac{\eta_{2\pi}(0)}{\Theta_{2\tau}\begin{bmatrix} 0 & 0 \\ \tfrac{1}{2} & 0 \end{bmatrix}(0)} = c$$

where $c_t(x)$ is the section of Prop. 5.1 and c is the constant of Prop. 4.1. Thus (120) implies Schottky's relations (80); the relations (85), on the other hand, are implied by (119) and Cor. 5.12: if α is an odd half period of the form $[\alpha] = \begin{bmatrix} 0 & \delta \\ 0 & \varepsilon \end{bmatrix}$ for $2\delta, 2\varepsilon \in (\mathbb{Z}/2\mathbb{Z})^{g-1}$, then

$$\lim_{t\to 0} G[\alpha](x,t) = \lim_{t\to 0} \sum_1^g \frac{\partial\eta_t[\alpha]}{\partial s_j}(0) w_j(x,t) = \sum_2^g \frac{\partial\eta\begin{bmatrix}\delta\\ \varepsilon\end{bmatrix}}{\partial s_j}(0) w_j(x) = G\begin{bmatrix}\delta\\ \varepsilon\end{bmatrix}(x)$$

provided x is kept away from the pinched region in $\hat{C}_0$; so by (121),

$$G\begin{bmatrix}\delta\\ \varepsilon\end{bmatrix}^2(x) = \lim_{t\to 0} G[\alpha]^2(x,t) = \lim_{t\to 0} \frac{\sigma_{\alpha,t}^2 H[\alpha]^2(x)\Theta[\alpha]^2(\int_{a_t}^{x} v - \tfrac{1}{2}\int_{a_t}^{b_t} v)}{\Theta[\alpha](\int_{a_t}^{x} v)\Theta[\alpha](\int_{b_t}^{x} v)}$$

$$= \frac{c^2 H[\alpha](p) H[\alpha]^2(x)\Theta\begin{bmatrix} 0 & \delta \\ \tfrac{1}{2} & \varepsilon \end{bmatrix}^2(\int_p^x v)}{H\begin{bmatrix} 0 & \delta \\ \tfrac{1}{2} & \varepsilon \end{bmatrix}(p)\Theta[\alpha](\int_p^x v)\Theta[\alpha](\int_p^x v)} = c^2 H\begin{bmatrix} 0 & \delta \\ 0 & \varepsilon \end{bmatrix}(x) H\begin{bmatrix} 0 & \delta \\ \tfrac{1}{2} & \varepsilon \end{bmatrix}(x).$$

Example 2. Let $\mathcal{C}$ be a family of curves of genus 2g over the unit t-disc whose fibers $\hat{C}_t$ are ramified double coverings of C with distinct branch points a_t and b_t for $t \neq 0$, while, at $t = 0$, the two branch points coalesce to a point $p \in C$ and the fiber $\hat{C}_0$ consists of two copies of C joined at p. We construct $\mathcal{C}$ by pinching a cycle homologous to zero as in §3, p. 37, in such a way that $\mathcal{C}$ is the analytic surface $x^2 - y^2 = t$ near p, a_t and b_t are the points $x = -\sqrt{t}$ and $x = \sqrt{t}$, and $p \in C$ is the point $x = y = t = 0$. The normalized differentials on $\hat{C}_t$ have expansions

$$\left.\begin{aligned} u_\alpha(x,t) &= v_\alpha(x) + \tfrac{1}{4}tv_\alpha(p)\omega(x,p) + o(t) \\ u_{\alpha'}(x,t) &= \tfrac{1}{4}tv_\alpha(p)\omega(x,p) + o(t) \end{aligned}\right\} \quad 1 \le \alpha \le g$$

for $x \in C$ outside the pinched region, where $v_1,\ldots,v_g$ are the normalized holomorphic differentials on C, $\omega(x,p)$ is the differential of the second kind on C with pole at $x = p$, and o(t) are holomorphic differentials on C outside the pinched region with $\lim_{t\to 0} \frac{1}{t} o(t) = 0$ there. The Riemann matrix for $\hat{C}_t$ is thus given by

$$\hat{\tau}(t) = \left(\begin{array}{c|c} \tau_{\alpha\beta} + \frac{t}{4} v_\alpha(p) v_\beta(p) & \frac{t}{4} v_\alpha(p) v_\beta(p) \\ \hline \frac{t}{4} v_\alpha(p) v_\beta(p) & \tau_{\alpha\beta} + \frac{t}{4} v_\alpha(p) v_\beta(p) \end{array}\right) + o(t) \qquad 1 \le \alpha,\beta \le g$$

where τ is the Riemann matrix for C and $\lim_{t\to 0} \frac{1}{t} o(t) = 0$. Using these expansions, we have

Proposition 5.13. For any $x,p \in C$, set $Z_x(p) = \frac{d}{dp} \ln E(x,p)$ as in Cor. 2.6 (iii); and, for $e \in \mathbb{C}^g$, let $\psi(z) = \Theta(z+x-p+e)\Theta(z-e)$, a second-order theta function on $\mathbb{C}^g$. Then

$$\sum_{i,j=1}^{g} v_i(p) v_j(p) \frac{\partial^2 \ln\psi}{\partial z_i \partial z_j}(0) + \left(\sum_{i=1}^{g} v_i(p) \frac{\partial \ln\psi}{\partial z_i}(0)\right)^2 + \sum_{i=1}^{g} \left(2 Z_x(p) v_i(p) + v_i'(p)\right) \frac{\partial \ln\psi}{\partial z_i}(0)$$

is a cochain in the sheaf of quadratic differentials in p, holomorphic for all x and $p \in C$ and independent of $e \in \mathbb{C}^g$.

<u>Proof</u>. When $x \in C$ is away from the point p, the expansions above give

$$\int_{a_t}^{x} w_\alpha(y,t) = \int_{a_t}^{x} u_\alpha(y,t) + u_{\alpha'}(y,t) = \int_p^x v_\alpha + \tfrac{1}{2}t(v_\alpha(p)Z_x(p) + d_\alpha) + o(t)$$

for some constants of integration d_α to be determined. On the other hand,

$$\int_{a_t}^{x} v - \tfrac{1}{2}\int_{a_t}^{b_t} v = \int_p^x v - \tfrac{1}{2}tv'(p) + o(t)$$

so that (97) of Prop. 5.1 for two branch points gives, $\forall\, e \in \mathbb{C}^g$:

$$\begin{aligned} c_t(x) = 1 + \frac{t}{8\,\theta(e)\,\theta(e+x-p)} \Bigg\{ & \theta(-e) \sum_{i,j=1}^{g} \frac{\partial^2\theta}{\partial z_i \partial z_j}(e+x-p)\, v_i(p) v_j(p) \\ & + 2\sum_{i=1}^{g} \frac{\partial\theta}{\partial z_i}(-e)\, v_i(p) \sum_{j=1}^{g} \frac{\partial\theta}{\partial z_j}(e+x-p)\, v_j(p) + \theta(e+x-p) \sum_{i,j=1}^{g} \frac{\partial^2\theta}{\partial z_i \partial z_j}(-e)\, v_i(p) v_j(p) \\ & + 2\sum_{i=1}^{g} \left(v_i'(p) + v_i(p) Z_x(p) + d_i\right)\left(\theta(-e)\frac{\partial\theta}{\partial z_i}(e+x-p) + \theta(e+x-p)\frac{\partial\theta}{\partial z_i}(-e)\right) \Bigg\} + o(t). \end{aligned}$$

The coefficient of t is independent of e and thus is holomorphic $\forall\, x,p \in C$; taking $e + \tfrac{1}{2}\int_p^x v = f \in (\Theta)$ non-singular, and equating to zero the coefficient of $\frac{1}{x-p}$ in the Laurent series at $x = p$, we conclude that $\sum_1^g (d_i + \tfrac{1}{2}v_i'(p))\frac{\partial\Theta}{\partial z_i}(f) = 0$ $\forall f$ which, by Cor. 4.21 implies that $d_i = -\tfrac{1}{2}v_i'(p)$.

VI. Bordered Riemann Surfaces

This chapter is an application of Θ-functions to the study of unitary functions, symmetric differentials and kernel functions on a finite bordered Riemann surface.

Let R be an open Riemann surface of genus ρ with a positively oriented boundary ∂R consisting of n disjoint analytic curves $\Gamma_0, \Gamma_1, \ldots, \Gamma_{n-1}$. The double C of $R \cup \partial R$ [3, p. 107] is a compact Riemann surface (without boundary) of genus $g = 2\rho+n-1$ admitting an anticonformal involution ϕ with fixed point set ∂R. We will choose a symmetric open covering of C by neighborhoods U_α with α in an index set $I \cup I_0 \cup I'$ as follows: for $\alpha \in I$ (resp. I'), $U_\alpha \subset R$ (resp. $\phi(R)$) will have local coordinate $z_\alpha : U_\alpha \to ₵$ such that $z_{\alpha'} = \overline{z_\alpha \circ \phi}$ is the local coordinate on an open set $U_{\alpha'} = \phi(U_\alpha)$ for some unique $\alpha' \in I'$ (resp. I); in case $\alpha \in I_0$, $U_\alpha = \phi(U_\alpha)$ is to be a symmetric boundary neighborhood having a local coordinate $z_\alpha : U_\alpha \to ₵$ real on ∂R and with positive imaginary part on $U_\alpha \cap R$. In terms of this covering the canonical bundle can be described by a cocycle $(k_{\alpha\beta}) = (\frac{dz_\beta}{dz_\alpha}) \in H^1(C, \mathcal{O}_C^*)$ with $k_{\alpha\beta}(x) > 0$ for $x \in \partial R$ and with symmetry $k_{\alpha\beta}(x) = \overline{k_{\alpha'\beta'}(\phi(x))}$ for $x \in U_\alpha \cap U_\beta$, so that v and $\overline{\phi^* v}$ are sections of the same bundle for any differential v on C.

Let us fix a symmetric canonical homology basis on C:

$$A_1, B_1, \ldots, A_\rho, B_\rho, A_{\rho+1}, B_{\rho+1}, \ldots, A_{\rho+n-1}, B_{\rho+n-1}, A_{1'}, B_{1'}, \ldots, A_{\rho'}, B_{\rho'}$$

such that $A_{\rho+k} = \Gamma_k$ for $k = 1, \ldots, n-1$, and $A_1, B_1, \ldots, A_\rho, B_\rho$ (resp. $A_{1'}, B_{1'}, \ldots, A_{\rho'}, B_{\rho'}$) are cycles in R (resp. $\phi(R)$) satisfying the relations in $H_1(C, \mathbb{Z})$:

$\phi(A_i) = A_{i'}, \quad \phi(B_i) = -B_{i'}, \quad 1 \le i \le \rho$

$\phi(A_i) = A_i, \quad \phi(B_i) = -B_i, \quad \rho+1 \le i \le \rho+n-1$

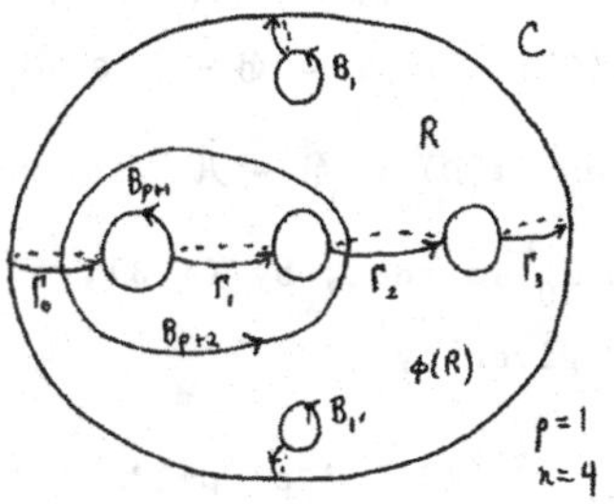

If $u_1,\dots,u_\rho,u_{\rho+1},\dots,u_{\rho+n-1},u_{1'},\dots,u_{\rho'}$ are the corresponding normalized differentials on C, then

(123) $\quad \phi^* u_i = -\bar{u}_{i'}, \quad 1 \le i \le \rho \quad$ and $\quad \phi^* u_i = -\bar{u}_i, \quad \rho+1 \le i \le \rho+n-1,$

and the period matrix for C has the symmetric form

$$\tau = \left(\int_{B_j} u_i\right) = \begin{pmatrix} a & b & c \\ b^t & d & \bar{b}^t \\ \bar{c} & \bar{b} & \bar{a} \end{pmatrix}$$

where a and c are $\rho \times \rho$ matrices, b is a $\rho \times n-1$ matrix and d is a *real* $(n-1)\times(n-1)$ matrix. The normalized differentials of the second and third kind on C have the symmetries

(124) $$\omega(x,y) = \overline{\omega(\bar{x},\bar{y})} \quad \text{and} \quad \omega_{b-a}(x) = \overline{\omega_{\bar{b}-\bar{a}}(\bar{x})}$$

for $x,y,a,b \in C$ and $\bar{x} = \phi(x)$ the conjugate point of $x \in C$. From (123), we also conclude that, for $b_0 \in \Gamma_0$ and $x \in C$,

(125)
$$\int_x^{\bar{x}} u = \begin{Bmatrix} m_1(x)..m_\rho(x) & m_{\rho+1}(x)..m_{\rho+n+1}(x) & m_1(x)..\,m_\rho(x) \\ n_1(x)..n_\rho(x) & 0 \quad .. \quad 0 & -n_1(x)..-n_\rho(x) \end{Bmatrix}_\tau \in \mathbb{C}^g$$

$$\int_{2b_0}^{x+\bar{x}} u = \begin{Bmatrix} \mathfrak{m}_1(x)..\mathfrak{m}_\rho(x) & 0 \quad .. \quad 0 & -\mathfrak{m}_1(x)..-\mathfrak{m}_\rho(x) \\ \mathfrak{n}_1(x)..\mathfrak{n}_\rho(x) & \mathfrak{n}_{\rho+1}(x)..\mathfrak{n}_{\rho+n-1}(x) & \mathfrak{n}_1(x)..\,\mathfrak{n}_\rho(x) \end{Bmatrix}_\tau \in \mathbb{C}^g$$

where the m, n, $\mathfrak{m}$, $\mathfrak{n}$ are the general harmonic measures on C, and $m_{\rho+1}(x),\dots,m_{\rho+n-1}(x)$ are well-defined functions bounded between 0 and 1 for $x \in R$.

The mapping ϕ gives rise to an antiholomorphic involution on $J_0(C)$: if $D = \mathcal{B} - \mathcal{A} \in J_0(C)$ with $\mathcal{A}, \mathcal{B}$ positive divisors on C then $\phi(D) = \overline{\mathcal{B}} - \overline{\mathcal{A}}$ is the class of the point $(\int_{\overline{\mathcal{A}}}^{\overline{\mathcal{B}}} u) = (\int_{\mathcal{A}}^{\mathcal{B}} \phi^* u)$ in J_0 so that, by (123), ϕ lifts to the antiholomorphic involution on $\mathbb{C}^g$ given by

$$\phi(z_1,\ldots,z_\rho,z_{\rho+1},\ldots,z_{\rho+n-1},z_{1'},\ldots,z_{\rho'}) = -(\bar{z}_{1'},\ldots,\bar{z}_{\rho'},\bar{z}_{\rho+1},\ldots,\bar{z}_{\rho+n-1},\bar{z}_1,\ldots,\bar{z}_\rho).$$

In terms of τ-characteristics of a point in $\mathbb{C}^g$, this equation becomes

$$(126) \qquad \phi\begin{Bmatrix}\alpha & \mu & \gamma\\ \beta & \nu & \delta\end{Bmatrix}_\tau = \begin{Bmatrix}-\gamma & -\mu & -\alpha\\ \delta & \nu & \beta\end{Bmatrix}_\tau \in \mathbb{C}^g$$

for all $\alpha,\beta,\gamma,\delta \in \mathbb{R}^\rho$ and $\mu,\nu \in \mathbb{R}^{n-1}$.

<u>Proposition 6.1.</u> The theta-function for J(C) formed from the period matrix τ has the symmetry

$$(127) \qquad \Theta\begin{bmatrix}\alpha & \mu & \gamma\\ \beta & \nu & \delta\end{bmatrix}(z) = \overline{\Theta\begin{bmatrix}-\gamma & -\mu & -\alpha\\ \delta & \nu & \beta\end{bmatrix}(\phi(z))} \qquad \forall\ z \in \mathbb{C}^g$$

for all $\alpha,\beta,\gamma,\delta \in \mathbb{R}^\rho$ and $\mu,\nu \in \mathbb{R}^{n-1}$. The Riemann divisor class $\Delta \in J_{g-1}(C)$ satisfies $\Delta = \phi(\Delta) \in J_{g-1}$, and the prime-form on $C \times C$ has the symmetry $E^2(x,y) = \overline{E(\bar{x},\bar{y})}^2\ \forall\ x,y \in C$.

<u>Proof.</u> By (1) and (126) it suffices to prove $\Theta(z) = \overline{\Theta(\phi(z))}$ to establish (127). But the quadratic form $Q(\xi) = \sum_{j,k=1}^{g} \xi_j\tau_{jk}\xi_k$ for $\xi \in \mathbb{R}^g$ has the symmetry $Q(\xi) = \overline{Q(\phi(\xi))}$ from the symmetry of τ; therefore

$$\overline{\Theta(\phi(z))} = \sum_{n\in\mathbb{Z}^g} \exp\{\tfrac{1}{2}n\bar{\tau}n^t + n\cdot\overline{\phi(z)}\} = \sum_{m\in\mathbb{Z}^g} \exp\{\tfrac{1}{2}\phi(m)\bar{\tau}\phi(m)^t + \phi(m)\cdot\overline{\phi(z)}\}$$

$$= \sum_{m\in\mathbb{Z}^g} \exp\{\tfrac{1}{2}m\tau m^t + m\cdot z\} = \Theta(z) \qquad \forall\ z \in \mathbb{C}^g.$$

To show that Δ is a symmetric point in $J_{g-1}(C)$, choose $a \in C$ and $e \in \mathbb{C}^g$ such that $\operatorname{div}_C \theta(x-a-e) = \mathcal{A} \not\geq C$; then by (127), $\operatorname{div}_C \theta(x-\bar{a}-\phi(e)) = \bar{\mathcal{A}}$ so that by Riemann's Theorem (10), $\bar{\mathcal{A}} - \bar{a} - \Delta = \phi(e) = \overline{\mathcal{A} - a - \Delta}$ and $\Delta = \bar{\Delta}$ in J_{g-1}. Finally, if f is any non-singular odd-half period (19) and (127) give:

$$\left(\sqrt{\frac{E(\bar{x},\bar{y})}{E(x,y)}}\right)^2 = \frac{\theta^2[\phi(f)](y-x)}{H[\phi(f)](\bar{x})H[\phi(f)](\bar{y})} \cdot \frac{H[f](x)H[f](y)}{\theta^2[f](y-x)} = 1$$

since the middle term is a well-defined meromorphic function on $C \times C$ with no zeroes or poles and with the value 1 along $y = x$.

Let us call a meromorphic function f on C *unitary* if $|f(x)| = 1$ for $x \in \partial R$; by the reflection principle all unitary functions are then, up to a constant, given by meromorphic functions f on C with $\operatorname{div}_C f = \bar{D} - D$ for some positive divisor D on $C - \partial R$.

Proposition 6.2. The subvariety $S = \{s \in J_0(C) \mid \phi(s) = s\}$ of J_0 is a disjoint union of 2^{n-1} real g-dimensional torii S_μ given by all points $\begin{Bmatrix} \alpha & \frac{1}{2}\mu & -\alpha \\ \beta & \nu & \beta \end{Bmatrix}_T \in J_0$ with $\alpha, \beta \in \mathbb{R}^\rho$, $\nu \in \mathbb{R}^{n-1}$ and $\mu = (\mu_1, \ldots, \mu_{n-1}) \in (\mathbb{Z}/2\mathbb{Z})^{n-1}$. If $\bar{D} - D$ is the divisor of a unitary function f on C and b is any point on Γ_0, then $D - (\deg D)b \in S_\mu$ with $\mu_k = \frac{1}{2\pi}\int_{\Gamma_k} d \arg f$ (modulo 2), for $k = 1, \ldots, n-1$.

Proof. If $s = \begin{Bmatrix} \alpha & \mu' & \gamma \\ \beta & \nu & \delta \end{Bmatrix}_T \in \mathbb{C}^g$ for $\alpha,\beta,\gamma,\delta \in \mathbb{R}^\rho$ and $\mu', \nu \in \mathbb{R}^{n-1}$, then by (126), $s - \phi(s) = \begin{Bmatrix} \alpha+\gamma & 2\mu' & \gamma+\alpha \\ \beta-\delta & 0 & \delta-\beta \end{Bmatrix}_T = 0$ in $J_0(C)$ if and only if $\alpha+\gamma \in \mathbb{Z}^\rho$, $\beta-\delta \in \mathbb{Z}^\rho$ and $2\mu' \in \mathbb{Z}^{n-1}$. Thus $S = \bigsqcup_{\mu \in (\mathbb{Z}/2\mathbb{Z})^{n-1}} S_\mu$, where S_μ is the set of all points in J_0 with characteristics $\begin{bmatrix} \alpha & \frac{1}{2}\mu & -\alpha \\ \beta & \nu & \beta \end{bmatrix}$ for $\alpha, \beta \in \mathbb{R}^\rho$, $\nu \in \mathbb{R}^{n-1}$ and $\mu = (\mu_1, \ldots, \mu_{n-1})$ with $\mu_i = 0$ or 1; S_0 is a real Abelian group of real dimension $2\rho+n-1 = g$

and S_μ, a translate of S_0 by the half-period $\begin{Bmatrix} 0 & \frac{1}{2}\mu & 0 \\ 0 & 0 & 0 \end{Bmatrix}_T$, is likewise a real torus of dimension g. Now suppose $s = D - (\deg D)b$ with $b \in \Gamma_0$ and $\bar{D} - D$ the divisor of a unitary function f on C. By (125) we can write

$$\phi(s) - s = \left(\int_D^{\bar{D}} u\right) = \begin{Bmatrix} m & -\mu & m \\ n & 0 & -n \end{Bmatrix}_T \in \mathbb{C}^g$$

with $m, n \in \mathbf{Z}^\rho$ and $\mu \in \mathbf{Z}^{n-1}$; applying Abel's theorem (8), we then conclude that $s \in S_\mu$ where $\mu_k = + \frac{1}{2\pi} \int_{\Gamma_k} d \arg f$ (modulo 2).

A divisor D on C is said to be *symmetric* if D is fixed under ϕ; symmetric divisors are thus of the form $D_1 + \bar{D}_1 + D_2$ where D_1 and D_2 are divisors on C with $D_2 \subset \partial R$. An alternate description of the torii S_μ is then given by

Proposition 6.3. For any fixed divisor $\mathcal{B}$ of degree g contained entirely in Γ_0, S_μ consists of all points $\mathcal{A} - \mathcal{B} \in J_0$ with $\mathcal{A}$ a positive symmetric divisor on C containing, modulo 2, μ_k points on Γ_k, $k = 1, \ldots, n-1$, and $n - 1 - \sum_1^{n-1} \mu_k$ points on Γ_0.

Proof. To prove the assertion first for $\mu = 0$, let Σ_0 be the set of all points $\mathcal{A} - \mathcal{B}$ with $\mathcal{A}$ a positive symmetric divisor containing an even number of points on each Γ_k, $k = 1, \ldots, n-1$; then $\Sigma_0 \subseteq S_0$ since by (123) and (125),

$$(128) \qquad \int_{2b_0}^{p+\bar{p}} u \in S_{0,\ldots,0} \quad \text{and} \quad \int_{b_0}^{b_k} u \in S_{0,\ldots,\underset{(k)}{1},\ldots,0}$$

for any $b_0 \in \Gamma_0$, $b_k \in \Gamma_k$ and $p \in C$. We will prove that $\Sigma_0 = S_0$ by showing that Σ_0 is an Abelian subgroup of S_0 containing some open neighborhood of real dimension g. Now for generic $x_1, \ldots, x_g \in \partial R$, the $g \times g$ matrix $(u_i(x_j))$ has non-vanishing determinant since otherwise there would be a holomorphic differential vanishing identically on ∂R, corresponding to the hyperplane in $\mathbb{P}_{g-1}(\mathbb{C})$ containing the image of ∂R under the canonical imbedding $x \in C \to [u_1(x),\dots,u_g(x)] \in \mathbb{P}_{g-1}(\mathbb{C})$. The Jacobian of the map $(x_1,\dots,x_g) \in (\partial R)^g \to x_1 + \dots + x_g - \mathcal{B} \in J_0$ has rank g at such generic points and consequently, the set Σ_0 contains an open neighborhood of real dimension g. On the other hand, suppose $\mathcal{A}_1 - \mathcal{B}$ and $\mathcal{A}_2 - \mathcal{B}$ are two points in Σ_0 and, using the Jacobi Inversion Theorem, let f be a meromorphic function on C with $\operatorname{div}_C f = X + \mathcal{B} - (\mathcal{A}_1 + \mathcal{A}_2)$ for X positive of degree g. Then $\operatorname{div}_C(f(x) \underset{(-)}{+} \overline{f(\bar{x})}) = \mathcal{A}_3 + \mathcal{B} - (\mathcal{A}_1 + \mathcal{A}_2)$ where $\mathcal{A}_3$ is a positive and symmetric divisor of degree g; thus $(\mathcal{A}_1 - \mathcal{B}) + (\mathcal{A}_2 - \mathcal{B}) = \mathcal{A}_3 - \mathcal{B} \in \Sigma_0$ and Σ_0 is a subgroup of S_0. From this last argument, we also see that a symmetric divisor of degree g is (linearly) equivalent to a positive symmetric divisor of the same degree, and this fact, together with (128) and the assertion for $\mu = 0$, proves the proposition for arbitrary μ.

Proposition 6.4. For any point $b \in \Gamma_0$, the vector of Riemann constants $k^b = \Delta - (g-1)b \in S_{1,1,\dots,1}$, so that for $s \in S_\mu$, $\Theta(x-b-s)$ either vanishes identically on C or has, modulo 2, $1 + \mu_k$ zeroes on Γ_k, $k = 1,\dots,n-1$ and $\sum_1^{n-1} \mu_k$ zeroes on Γ_0.

Proof. For $1 \le j \le g$, the jth component k_j^b of $k^b = \Delta - (g-1)b$ is given by (13); from (123) and the symmetry of τ:

$$\bar{k}^b_{j'} = \frac{-\bar{\tau}_{j'j'} - 2\pi i}{2} - \frac{1}{2\pi i}\sum_{\substack{k=1\\ k\neq j'}}^{g} \int_{A_k} \bar{u}_k \int_b^x \bar{u}_{j'}$$

$$(129) \qquad = \frac{-\tau_{jj} - 2\pi i}{2} - \frac{1}{2\pi i}\left\{\sum_{k\neq j'} \int_{A_{k'}} u_{k'} \int_b^x u_j + 2\pi i \int_{B_{k'}} u_j\right\}$$

$$= \frac{\tau_{jj} - 2\pi i}{2} - \frac{1}{2\pi i}\sum_{\ell \neq j} \int_{A_\ell} u_\ell \int_b^x u_j - \sum_{\ell=1}^{g} \tau_{\ell j}$$

where, as in Prop. 1.2, the paths of integration A_k are considered part of the *positively* oriented boundary of C dissected along generators of $\pi_1(C,b)$. Thus $k^b - \phi(k^b) = -\begin{Bmatrix} 1 & 1 & \dots & 1 & 1 \\ 0 & 0 & \dots & 0 & 0 \end{Bmatrix}_T \in \mathbb{C}^g$ and $k^b \in S_{1,\dots,1}$ from (126) since $k^b \in S_\mu$ for some $\mu \in (\mathbb{Z}/2\mathbb{Z})^{n-1}$ by Prop. 6.1. Now suppose that $s = \mathcal{A} - b - \Delta \in S_\mu$ with $b \in \Gamma_0$ and $\mathcal{A} = \mathrm{div}_C \theta(x-b-s) \neq C$; setting $\mathcal{A} = \mathcal{A}_1 + \overline{\mathcal{A}}_1 + \mathcal{B}$ with $\mathcal{B} \subset \partial R$ by Prop. 6.1, (125) implies that

$$s = \int_{gb}^{\mathcal{A}_1 + \overline{\mathcal{A}}_1 + \mathcal{B}} u - k^b = \int_{(\deg \mathcal{B})b}^{\mathcal{B}} u + \begin{Bmatrix} * & \frac{1}{2} & \dots & \frac{1}{2} & * \\ * & * & \dots & * & * \end{Bmatrix}_T \in S_{\mu_1,\dots,\mu_{n-1}}.$$

This means, by (128), that $\mathcal{B}$ must have (modulo 2) $1 + \mu_k$ points on Γ_k; and since $\deg \mathcal{A} = g = 2\rho + n - 1 \equiv n-1 \pmod 2$, $\mathcal{B}$ must have $\sum_1^{n-1} \mu_k \pmod 2$ points on Γ_0.

Corollary 6.5. $(\theta) \cap S$ is the $g-1$ real-dimensional variety consisting of points $s = \zeta - \Delta \in J_0$ with ζ a positive symmetric divisor of degree $g-1$ having (mod 2) $1 + \mu_k$ points on Γ_k if $s \in S_\mu$.*

Proof. If $s \in \theta \cap S_\mu$ and $b \in \Gamma_0$, then by Prop. 6.3, $s = \mathcal{A} - b - \Delta$ for some positive symmetric divisor $\mathcal{A}$. If $b \in \mathcal{A}$, $\mathcal{A} - b$ is a symmetric positive divisor ζ; if $b \notin \mathcal{A}$, $b + \zeta_1 - \mathcal{A}$ is the divisor of a meromorphic function f for some positive divisor ζ_1, and then $\mathrm{div}_C(f(x) \underset{(-)}{+} \overline{f(\bar{x})}) = b + \zeta - \mathcal{A} = 0$ in J_0 for some positive symmetric divisor ζ which, by Prop. 6.4, must have (modulo 2) $1 + \mu_k$ points on Γ_k.

Let $\hat{S}_\mu \subset \mathbb{C}^g$ be the universal cover of $S_\mu \subset J_0$ passing through the half-period $\begin{Bmatrix} 0 & \frac{1}{2}\mu & 0 \\ 0 & 0 & 0 \end{Bmatrix}_T \in \mathbb{C}^g$; then $\hat{S}_\mu$ is given by all points $\begin{Bmatrix} \alpha & \frac{1}{2}\mu & -\alpha \\ \beta & \nu & \beta \end{Bmatrix}_T \in \mathbb{C}^g$ with $\alpha, \beta \in \mathbb{R}^\rho$ and $\nu \in \mathbb{R}^{n-1}$, and by (126-7) $\theta[s](0)$ is real for all $s \in \hat{S}_\mu$. The spaces S_μ parametrize the generic unitary functions on C with the minimal (g+1) number of zeroes:

---

* Obviously $(\theta) \cap S_\mu \neq \emptyset$ for all $\mu \neq 0$; but $(\theta) \cap S_0$ can be empty - see p. 128.

<u>Proposition 6.6</u>. For any fixed $a \in C - \partial R$, let Σ_a be the subvariety $(S \cap \Theta) \cup V_a \cup W_a \subset S$ where

$$V_a = \bigsqcup_{b \in \partial R} \{b-a-\bar{a}+\mathcal{B}-\Delta \mid \mathcal{B} \text{ positive symmetric of degree } g \text{ and } i(\mathcal{B}) > 0\}$$

and

$$W_a = \bigsqcup_{\sigma} \{\sigma-a-\bar{a}+\Theta_{sing} \cap S \mid \sigma \text{ positive symmetric of degree 2 on } C\}.$$

Then for $s \in \hat{S}_\mu - \hat{S}_\mu \cap (\Theta)$,

$$(130) \qquad f(x) = \varepsilon \frac{\theta(x-\bar{a}-s)}{\theta(x-a-s)} \frac{E(x,a)}{E(x,\bar{a})} \exp \tfrac{1}{2} \sum_1^{n-1} \mu_k \int_a^{\bar{a}} u_{\rho+k}, \qquad |\varepsilon| = 1$$

is a unitary function on C vanishing at a, with $\frac{1}{2\pi} \int_{\Gamma_k} d \arg f = 1 + \mu_k$ (modulo 2) and with at most g (resp. g-1) poles if $s \in V_a$ (resp. W_a). Every unitary function f on C with a divisor D of exactly $g+1$ poles satisfying $i(D) = 0$ has the form (130) for some $\bar{a} \in D$ and $s \in \hat{S}_\mu - \hat{S}_\mu \cap \Sigma_a$.

<u>Proof</u>. If $s \in \hat{S}_\mu - \hat{S}_\mu \cap (\Theta)$, $\frac{\theta(x-\bar{a}-s)}{\theta(x-a-s)} \frac{E(x,a)}{E(x,\bar{a})}$ is a meromorphic function on C vanishing at a, with change in argument along Γ_k given by Props. 6.2 and 6.4, and with constant absolute value $\exp \frac{1}{2} \sum_1^{n-1} \mu_k \int_{\bar{a}}^{a} u_{\rho+k}$ on ∂R since $|E(b,a)| = |E(b,\bar{a})|$ for $b \in \Gamma_0$ and

$$|\theta(b-\bar{a}-s)| = |\theta(b-a-\phi(s))| = |\theta(b-a-s)| \exp \operatorname{Re} \sum_1^{n-1} \mu_k \int_b^a u_{\rho+k}$$

by Prop. 6.1. Thus f(x) as defined by (130) is a unitary function on C with at most $g+1$ poles; and f(x) will have a divisor of poles of degree $< g+1$ in two possible cases. First, $\operatorname{div}_C \theta(x-a-s)$ could be of the form $b+\zeta$ for some $b \in \partial R$ and ζ positive of degree $g-1$; in this case $\zeta+\bar{a} = \bar{\zeta}+a \in J_g$ is a special divisor of degree g since $a \notin \zeta$, and $s = b-a-\bar{a}+\mathcal{B}-\Delta \in V_a$ where $\mathcal{B} = \zeta+\bar{a}$ may be taken positive by Prop. 6.2. The other possibility is that $\text{div}_C\Theta(x-a-s) = c+\bar{c}+\eta$ for some $c \in C$ and η positive of degree $g-2$; then f has at most $g-1$ poles and $\eta+\bar{a} = \bar{\eta}+a \in J_{g-1}$ is a special divisor of degree $g-1$, implying that $s = c+\bar{c}-a-\bar{a}+(\eta+\bar{a}-\Delta) \in W_a$. Conversely, such a choice of $s \in V_a$ (resp. W_a) obviously gives rise to a special unitary function (130) with at most g (resp. g-1) poles. Finally, if a unitary function f(x) has a divisor D of $g+1$ poles satisfying $i(D) = 0$, then $i(D-a) = 0$ for any $\bar{a} \in D$ and by Riemann's Theorem, $\Theta(x-\bar{a}-s) = \Theta(D-a-x-\Delta) \not\equiv 0$ on C for $s = D-a-\bar{a}-\Delta \in S$; thus f(x) has the form (130) for some $|\varepsilon| = 1$ and $s \in \hat{S}_\mu - \hat{S}_\mu \cap \Sigma_a$.

<u>Remarks</u>. The variety V_a is at most $g-1$ real-dimensional since $V_a = \bigsqcup_{b\in\partial R} V_a^b$ where $V_a^b = \{-s \in S \mid s = \xi+a+\bar{a}-b-\Delta,\ \xi$ a positive symmetric divisor of degree $g-2\}$. In the general case, when f(x) has $g+1$ zeroes including one at a, the number of zeroes or poles of f in R will be constant on each of the components of the complement in S of the $g-1$ real-dimensional variety $V_a \sqcup (S \cap \Theta)$. It has been proved in [3, p. 126] that there are always unitary functions with exactly $g+1$ zeroes *all* in R; and when R is a planar domain, it is shown in Prop. 6.16 that $S_{0,\ldots,0} \cap \Sigma_a$ is empty for $a \in R$ and that the unitary functions holomorphic on R with $g+1$ zeroes are parametrized by the torus S_0.

Now suppose f is a unitary function on C with a divisor D of $g+1$ poles satisfying $i(D) = 0$; then $f_\lambda(x) = \dfrac{f(x) - \lambda}{1 - \bar{\lambda}f(x)}$, for $|\lambda| \neq 1$, is a unitary function with a divisor of $g+1$ poles $D_\lambda = D \in J_{g+1}$. If f is given by (130) for $s = D-a-\bar{a}-\Delta$ and $\bar{a} \in D$, then f_λ is given by (130) for $\bar{a}$ replaced by $\bar{c} \in D_\lambda$ and s by $s_\lambda = s - \int_{a+\bar{a}}^{c+\bar{c}} u$, where $\lambda = f(c)$, $|\lambda| \neq 1$; observe that $s_\lambda \notin V_c \cup W_c$ since V_c and W_c are translates of V_a and W_a by $\int_{c+\bar{c}}^{a+\bar{a}} u$, and $s_\lambda \notin (\Theta)$ since $i(D-c-\bar{c}) > 0$

for $c \in C$ iff $|\lambda| = |f(c)| = 1$. All functions f_λ have the same ramification points $df_\lambda = 0$ and locus $|f_\lambda| = 1$ given by

Corollary 6.7. If f is a unitary function on C with a divisor D of g+1 poles satisfying $i(D) = 0$, then the symmetric divisor of 4g zeroes of the differential $d\ln f$ - that is, the ramification points of the covering $f\colon C \to \mathbb{P}_1(\mathbb{C})$ - are given by $\operatorname{div}_C \Theta(2x+\Delta-D)$, a fourth order theta function on C by (2). The curves $|f(x)| = 1$ on C lying over the unit circle in $\mathbb{P}_1(\mathbb{C})$ are the components of ∂R together with the locus $\Theta(x+\bar{x}+\Delta-D) = 0$.*

Proof. By Prop. 6.6, f has the form (130), so (38)′ implies

$$d\ln f = d\ln \frac{\Theta(x-\bar{a}-s)}{\Theta(x-a-s)}\frac{E(x,a)}{E(x,\bar{a})} = \frac{\Theta(s)\Theta(2x-a-\bar{a}-s)\,E(\bar{a},a)}{\Theta(x-a-s)\Theta(x-\bar{a}-s)E(x,a)E(x,\bar{a})}$$

which gives the first assertion since $s = D-\bar{a}-a-\Delta$. On the other hand, the addition theorem (45) gives

$$f(x) - f(\bar{x}) = \varepsilon\,\frac{\Theta(x+\bar{x}-a-\bar{a}-s)\Theta(s)E(x,\bar{x})E(\bar{a},a)}{\Theta(x-a-s)\Theta(x-\bar{a}-s)E(x,a)E(x,\bar{a})}\,\exp \tfrac{1}{2}\sum_{k=1}^{n-1}\mu_k\int_a^{\bar{a}} u_{\rho+k}$$

and thus the zeroes of the harmonic function $f(x) - f(\bar{x})$, describing the locus $|f(x)| = 1$, are $\operatorname{div}_C \Theta(x+\bar{x}+\Delta-D)E(x,\bar{x})$.

We say that a meromorphic differential v on C is *symmetric* if $v = \overline{\phi^* v}$ or equivalently, if $v(x) = \lambda w(x)$ for a differential w satisfying $\operatorname{div}_C w = \overline{\operatorname{div}_C w}$ and for a suitable constant λ depending on w. In terms of the symmetric boundary coordinates given on p. 108, such a differential v is then real on ∂R, and the sign of v at a point of ∂R (not a zero of v) is well-defined since the canonical cocycle $(k_{\alpha\beta})$ is positive on ∂R. A symmetric differential will be called *definite* if it does not change sign along each contour $\Gamma_0, \Gamma_1, \ldots, \Gamma_{n-1}$ - that is, if all its zeroes or poles on ∂R occur with even order.

---

* This can be empty - see Prop. 6.16.

Proposition 6.8. The subvariety $T = \{t \in J_0(C) \mid \phi(t) = -t\}$ of J_0 is a disjoint union of the 2^{n-1} real g-dimensional torii T_ν given by the points $\begin{Bmatrix} \alpha & \mu & \alpha \\ \beta & \frac{1}{2}\nu & -\beta \end{Bmatrix}_\tau \in J_0$, with $\alpha,\beta \in \mathbb{R}^\rho$, $\mu \in \mathbb{R}^{n-1}$ and $\nu = (\nu_1,\ldots,\nu_{n-1}) \in (\mathbb{Z}/2\mathbb{Z})^{n-1}$. Each torus T_ν consists of all points $t = D - \Delta \in J_0$ with $D + \bar{D}$ the divisor of a definite symmetric differential on C, holomorphic if $t \in (\Theta)$, non-negative on Γ_0 and real with sign $(-1)^{\nu_k}$ along Γ_k, $k = 1,\ldots,n-1$.

Proof. Let $t = \begin{Bmatrix} \alpha & \mu & \gamma \\ \beta & \nu' & \delta \end{Bmatrix}_\tau \in T$ for $\alpha,\beta,\gamma,\delta \in \mathbb{R}^\rho$ and $\mu,\nu' \in \mathbb{R}^{n-1}$; then by (126), $t + \phi(t) = \begin{Bmatrix} \alpha-\gamma & 0 & \gamma-\alpha \\ \beta+\delta & 2\nu' & \delta+\beta \end{Bmatrix}_\tau = 0$ in $J_0(C)$ if and only if $\alpha-\gamma \in \mathbb{Z}^\rho$, $\beta+\delta \in \mathbb{Z}^\rho$ and $2\nu' \in \mathbb{Z}^{n-1}$. Thus $T = \bigsqcup_{\nu \in (\mathbb{Z}/2\mathbb{Z})^{n-1}} T_\nu$ where T_ν is the set of all points in J_0 with characteristics $\begin{bmatrix} \alpha & \mu & \alpha \\ \beta & \frac{1}{2}\nu & -\beta \end{bmatrix}$, $\nu \in \mathbb{Z}^{n-1}$, a translate by the half-period $\begin{Bmatrix} 0 & 0 & 0 \\ 0 & \frac{1}{2}\nu & 0 \end{Bmatrix}_\tau$ of the group T_0 of real dimension g. Now by the Jacobi Inversion Theorem, any $t \in J_0$ can be written as $t = D - \Delta$ for D of degree g-1 and, by Prop. 6.1, $D - \Delta + \bar{D} - \Delta = 0 \in J_0$ if $t \in T$; this means that $D + \bar{D}$ is the divisor of a symmetric differential on C, definite on ∂R since all zeroes occur to even order on ∂R. In order to determine the appropriate sign arrangements, suppose $t_1 = D_1 - \Delta \in T_{\nu^1}$ and $t_2 = D_2 - \Delta \in T_{\nu^2}$ for divisors D_1 and D_2 on C; then the ratio of the two symmetric differentials corresponding to t_1 and t_2 will be a multiple of the symmetric function $\exp\left\{\int_b^x \omega_{D_1+\bar{D}_1-D_2-\bar{D}_2} + \sum_{j=1}^{\rho} m_j \int_b^x (u_j - u_{j'})\right\}$ for $b \in \Gamma_0$, $m \in \mathbb{Z}^\rho$, and $\begin{Bmatrix} m & 0 & -m \\ * & * & * \end{Bmatrix}_\tau = \int_{D_1+\bar{D}_1}^{D_2+\bar{D}_2} u \in \mathbb{C}^g$.

This function is positive for $x \in \Gamma_0$ by (124) and real with sign along Γ_k given by $(-1)^{\varepsilon_k}$ where, for any $b_k \in \Gamma_k$,

$$\varepsilon_k = \frac{1}{\pi} \arg \exp\left\{\int_b^{b_k} \omega_{D_1+\bar{D}_1-D_2-\bar{D}_2} + \sum_1^{\rho} m_j \int_b^{b_k} (u_j - u_{j'})\right\}$$

$$= \frac{1}{\pi}\,\mathrm{Im}\left\{\int_b^{b_k}\omega_{D_1-D_2} + \int_b^{\bar{b}_k}\bar{\omega}_{D_1-D_2} + \sum_1^{\rho} m_j\Big(\int_b^{b_k}u_j + \int_b^{\bar{b}_k}\bar{u}_j\Big)\right\}$$

$$= \frac{1}{\pi}\{\mathrm{Im}\int_{D_1}^{D_2}u_k - \sum_1^{\rho} m_j\,\mathrm{Im}\,\tau_{jk}\} = \nu_k^2 - \nu_k^1 \text{ (modulo 2)}$$

by (7) and the symmetries (123-4). Thus two symmetric definite differentials arise from points in distinct torii T_ν if and only if they have a different sign arrangement along ∂R. Now there are points in all torii, except possibly $T_{0,0,\ldots,0}$, giving rise to holomorphic definite differentials since for any $\nu \neq 0$, there is always an odd half-period in T_ν making $(\Theta) \cap T_\nu$ non-empty. But by Cauchy's Theorem, there are no holomorphic symmetric differentials non-negative everywhere on ∂R; thus $T_{0,\ldots,0}$ must be the torus giving rise to the differentials non-negative on ∂R and always *meromorphic*.

Let $\hat{T}_\nu \subset \mathbb{C}^g$ be the universal cover of $T_\nu \subset J_0$ passing through the half-period $\begin{Bmatrix} 0 & 0 & 0 \\ 0 & \frac{1}{2}\nu & 0 \end{Bmatrix}_\tau \in \mathbb{C}^g$; then $\hat{T}_\nu$ is given by all points $\begin{Bmatrix} \alpha & \mu & \alpha \\ \beta & \frac{1}{2}\nu & -\beta \end{Bmatrix}_\tau \in \mathbb{C}^g$ with $\alpha,\beta \in \mathbb{R}^\rho$ and $\mu \in \mathbb{R}^{n-1}$, and by Prop. 6.1, $\Theta(t)$ is real for all $t \in \hat{T}_\nu$.

<u>Corollary 6.9</u>. If $t \in \hat{T}_\nu$, $\dfrac{\Theta(x-a-t)\Theta(x-\bar{a}+t)}{E(x,a)E(x,\bar{a})}$ is a symmetric differential on C, holomorphic if $t \in (\Theta)$ and real with sign $(-1)^{\nu_k}$ along Γ_k, $k = 0,1,\ldots,n-1$ (with the convention $\nu_0 = 0$). The bilinear differential $\dfrac{\Theta(y-x-t)\Theta(y-x+t)}{E^2(x,y)}$ is real with sign $(-1)^{\nu_k+\nu_\ell}$ whenever $x \in \Gamma_k$ and $y \in \Gamma_\ell$, $0 \le k,\ell \le n-1$.

<u>Proof</u>. Since $t + \phi(t) = \begin{Bmatrix} 0 & 0 & 0 \\ 0 & \nu & 0 \end{Bmatrix}_T \in \mathbb{C}^g$, (2) and Prop. 6.1 imply

$$\frac{\Theta(b-a-t)\Theta(b-\bar{a}+t)}{E(b,a)E(b,\bar{a})} = \frac{\Theta(b-a-t)}{E(b,a)}\;\frac{\Theta(\bar{b}-\bar{a}-\phi(t)+2\pi i\nu)}{\overline{E(b,a)}} = \left|\frac{\Theta(b-a-t)}{E(b,a)}\right|^2 \ge 0$$

for any $b \in \Gamma_0$ and $a \in C$ near b; from continuity in a then, the symmetric definite differential $\dfrac{\Theta(x-a-t)\Theta(x-\bar{a}+t)}{E(x,a)E(x,\bar{a})} \geq 0$ for $x \in \Gamma_0$, which gives the first assertion by Prop. 6.8. The second assertion comes from setting $a = y \in \Gamma_\ell$, since we have just seen that the sign of $\dfrac{\Theta(b-y-t)\Theta(b-y+t)}{E(b,y)E(b,y)}$ is $(-1)^{\nu_\ell}$ for $b \in \Gamma_0$.

This corollary, together with (25) and (39), implies that $\forall\, x \in \Gamma_k$ and $y \in \Gamma_\ell$, $k,\ell = 0,\ldots,n-1$,

$$(-1)^{\nu_k+\nu_\ell} H_f(x)H_f(y) \leq 0$$

for any non-singular point $f \in \hat{T}_\nu \cap (\Theta)$, and

$$(-1)^{\nu_k+\nu_\ell}\left[\omega(x,y) + \sum_{i,j=1}^{g} \frac{\partial^2 \ln \Theta}{\partial z_i \partial z_j}(t)u_i(x)u_j(y)\right] \geq 0$$

for any $t \in \hat{T}_\nu$ with $\Theta(t) \neq 0$. Integrations of these bilinear differentials over $y \in \Gamma_\ell$ and $x \in \Gamma_k$ give holomorphic differentials with prescribed signs along ∂R, as well as various inequalities for the partial derivatives of Θ. From Prop. 6.4 and Cor. 6.9, one also concludes

Corollary 6.10. For each $\mu,\nu \in (\mathbb{Z}/2\mathbb{Z})^{n-1}$, $S_\mu \cap T_\nu$ consists of 4^ρ symmetric half-periods of the form $\begin{Bmatrix} \delta & \frac{1}{2}\mu & \delta \\ \varepsilon & \frac{1}{2}\nu & \varepsilon \end{Bmatrix}_\tau$, 2δ and $2\varepsilon \in (\mathbb{Z}/2\mathbb{Z})^\rho$. If $e \in S_\mu \cap T_\nu$ and $b \in \Gamma_0$, the half-order differential $\dfrac{\Theta[e](x-b)}{E(x,b)}$ either vanishes identically on C or has $1+\mu_k$ (modulo 2) zeroes on Γ_k and is real on Γ_0 and real (resp. imaginary) on Γ_k for $\nu_k = 0$ (resp. 1).

The transition functions defining the corresponding bundles L_e of half-order differentials can be found from the following

Proposition 6.11. In terms of the symmetric open cover $\{U_\alpha\}$ of C described on p. 108, the bundle of half-order differentials L_e can be given by a cocycle $(g_{\alpha\beta}) \in H^1(C,\mathcal{O}_C^*)$ with $g_{\alpha\beta}^2(x) = k_{\alpha\beta}(x)$ and $g_{\alpha\beta}(x) = \overline{g_{\alpha'\beta'}(\bar{x})}$ if and only if e is one of the 2^g half periods in T_0.

Proof. Any cocycle $(e_{\alpha\beta}) \in H^1(C,\mathbb{C}^*)$ will satisfy $e_{\alpha\beta}^2 = 1$ and $e_{\alpha\beta}(x) = \overline{e_{\alpha'\beta'}(\bar{x})}$ if and only if $(e_{\alpha\beta})$ corresponds to a line bundle of the form $e = \begin{Bmatrix} \delta_1 & \mu & \delta_1 \\ \delta_2 & 0 & \delta_2 \end{Bmatrix} \in T_0 \cap S$ since $-e = e = \phi(e) \in J_0(C)$ and the characteristic homomorphism of $(e_{\alpha\beta})$ over the cycle $B_{\rho+j}$, $j = 1,\ldots,n-1$, is $\prod_{k=1}^{N_j} e_{i_{k-1}i_k} e_{i'_{k-1}i'_k} = \prod_{k=1}^{N_j} e^2_{i_{k-1}i_k} = 1$ for a chain $U_{i_0},\ldots,U_{i_{N_j}}$ of neighborhoods in R joining some boundary neighborhoods U_{i_0} and $U_{i_{N_j}}$ for Γ_0 and Γ_j, respectively. Therefore, by a standard construction of a cocycle from the characteristic homomorphism of a line bundle - see [13, p. 186] - it will suffice to prove only that L_0 can be described by a cocycle of the form $(g_{\alpha\beta}) \in H^1(C,\mathcal{O}^*)$ with $g_{\alpha\beta}^2 = k_{\alpha\beta}$ and $g_{\alpha\beta}(x) = \overline{g_{\alpha'\beta'}(\bar{x})}$. So suppose L_0 is given by the cocycle $(g_{\alpha\beta})$ with $g_{\alpha\beta}^2 = k_{\alpha\beta}$ and set $\varepsilon_{\alpha\beta}(x) = \dfrac{\overline{g_{\alpha'\beta'}(\bar{x})}}{g_{\alpha\beta}(x)}$; then $\varepsilon_{\alpha\beta}^2 = 1$ since $\overline{k_{\alpha'\beta'}(\bar{x})} = k_{\alpha\beta}(x)$, and $(\varepsilon_{\alpha\beta})$ is a trivial cocycle in $H^1(C,\mathbb{C}^*)$ since $\phi^*\bar{L}_0 \cong L_0$ by Prop. 6.1. With no loss of generality we can assume that $\varepsilon_{\alpha\beta} = 1$ whenever α,β are in the index set I_0 (see p. 108) since $k_{\alpha\beta}(x) > 0$ for $x \in \partial R$ with the positive orientation; we will then be finished if it can be shown that $\tilde{\varepsilon}_{\alpha\beta} = \begin{cases} 1 & \alpha,\beta \in I_0 \cup I' \\ \varepsilon_{\alpha\beta} & \alpha \in I, \quad \beta \in I_0 \cup I \end{cases}$ defines a trivial cocycle $(\tilde{\varepsilon}_{\alpha\beta}) \in H^1(C,\mathbb{C}^*)$. But for $b \in \Gamma_0$, $\dfrac{\theta(x-b)}{E(x,b)} \not\equiv 0$ is a section of L_0 which is real on ∂R since its square is a positive differential on ∂R by Prop. 6.8; this means that if $g_\alpha(x)$ is the section $\dfrac{\theta(x-b)}{E(x,b)}$ on U_α, $g_\alpha(x) = g_{\alpha\beta}(x)g_\beta(x)$ for $x \in U_\alpha \cap U_\beta$ and $\dfrac{\overline{g_{\alpha'}(\bar{x})}}{g_\alpha(x)} = \varepsilon_\alpha(x) = \pm 1$ for $x \in U_\alpha$, where $\varepsilon_\alpha = 1$ if U_α $(\alpha \in I_0)$

is a boundary neighborhood. Consequently, if $\alpha \in I$ and $\beta \in I \cup I_0$, $\tilde{\varepsilon}_{\alpha\beta}(x) = \varepsilon_{\alpha\beta}(x) = \frac{\varepsilon_\alpha}{\varepsilon_\beta}(x)$ for $x \in U_\alpha \cap U_\beta$, and the cocycle $(\tilde{\varepsilon}_{\alpha\beta})$ is therefore trivial on C.

<u>Corollary 6.12</u>. The prime form on $C \times C$ has the symmetries

$$E(x,y) = \overline{E(\bar{x},\bar{y})} \quad \text{and} \quad E(b,x) = \overline{E(b,\bar{x})} \tag{131}$$

for all $x,y \in C$, $b \in \Gamma_0$. If $\delta(p) = \bar{p} - p \in J_0(C)$ for $p \in C$, and if $|K_C|$ is the real line bundle defined by the cocycle $|k_{\alpha\beta}|$, $iE(p,\bar{p})$ is a real C^∞-section of $|K_C|^{-1} \otimes \delta^*(\Theta)$ which is strictly positive (resp. negative) for $p \in R$ (resp. $\bar{R}$).

<u>Proof</u>. By Prop. 6.1, $(\overline{E(\bar{x},\bar{y})}/E(x,y))^2$ is the constant function 1 on $C \times C$. Therefore since $\overline{E(\bar{x},\bar{y})}/E(x,y)$, a section of $\phi^*\bar{L}_0^{-1} \otimes L_0$ in x and y, is a well-defined *function* on $C \times C$ by Prop. 6.11, it must actually be the constant given by $\lim_{x,y \to b \in \Gamma_0} \overline{E(\bar{x},\bar{y})}/E(x,y) = 1$. Taking $x = p$ and $y = \bar{p}$, $iE(p,\bar{p}) = -iE(\bar{p},p)$ defines by (131) a real C^∞-section of $L_0^{-1} \otimes \phi^*\bar{L}_0^{-1} \otimes \delta^*(\Theta)$, a bundle with positive transition functions by Prop. 6.11 and the fact that $\Theta(\bar{p} - p)$ picks up the factor $\exp \mathrm{Re}\left\{-\tau_{jj} - \tau_{jj'} + 2\int_p^{\bar{p}} u_j\right\} \in \mathbb{R}^+$ (resp. 1) as p describes the loop B_j (resp. A_j). For $p \in R$, $iE(p,\bar{p})$ is never zero and is strictly positive since the transition functions are positive and $iE(p,\bar{p})|dz_0(p)| = 2 \,\mathrm{Im}\, z_0(p) + \ldots > 0$ if $p \in R$ is near a point $p_0 \in \Gamma_0$ with a symmetric boundary coordinate z_0.

For a suitable choice of homology basis on a planar domain R, $1/iE(p,\bar{p})$ is called the capacity (or transfinite diameter) of R with respect to the point p - see (133).

<u>Corollary 6.13</u>. For all $t \in \hat{T}_0$, $\Theta(t) > 0$ and $\dfrac{\Theta(t-a+\bar{a})}{i\Theta(t)E(a,\bar{a})}$ is a section of $|K_C| \otimes 2\,\mathrm{Re}\, t$, strictly positive (resp. negative) for $a \in R$ (resp. $\bar{R}$).

Proof. By Prop. 6.1, $\theta(t) \in \mathbb{R}\ \forall t \in \hat{T}_0$ and by Prop. 6.8, $\theta(t)$ is never zero for any $t \in T_0$. But the sign of $\theta(0)$ is a continuous function of the moduli and so must remain constant as R is pinched along a loop enclosing ∂R as in §3. From Cor. 3.2 and the symmetry of τ, the limiting value of $\theta(0)$ is the positive quantity

$$\sum_{\substack{n_1,n_2 \in \mathbb{Z}^\rho \\ m \in \mathbb{Z}^{n-1}}} \exp \tfrac{1}{2}\{n_1 a n_1^t + n_2 \bar{a} n_2^t + m d m^t\} = \Big|\sum_{n \in \mathbb{Z}^\rho} e^{\frac{1}{2} n a n^t}\Big|^2 \sum_{m \in \mathbb{Z}^{n-1}} e^{\frac{1}{2} m d m^t}$$

where d is the real period matrix of a planar domain, and a is the period matrix of a compact Riemann surface of genus ρ which may be assumed generic - that is, $\theta_a(0) \neq 0$. The assertion concerning $\dfrac{\theta(t-a+\bar{a})}{i\theta(t)E(a,\bar{a})}$ then follows from the property of $iE(a,\bar{a})$ given in Cor. 6.12, and is also a direct consequence of Cor. 6.9:

$$\frac{\theta(t-a+\bar{a})}{i\theta(t)E(a,\bar{a})} = i \operatorname*{Res}_{x=a} \frac{\theta(x-\bar{a}-t)\theta(x-a+t)}{\theta^2(t)E(x,a)E(x,\bar{a})} = \frac{1}{2\pi\theta^2(t)} \int_{\partial R} \frac{\theta(x-\bar{a}-t)\theta(x-a+t)}{E(x,a)E(x,\bar{a})} > 0.$$

When $t = \begin{Bmatrix} 0 & \mu & 0 \\ \beta & 0 & -\beta \end{Bmatrix}_\tau \in \hat{T}_0$ with $\mu \in \mathbb{R}^{n-1}$ and $\beta \in \mathbb{R}^\rho$, $\dfrac{\theta(t-a+\bar{a})}{i\theta(t)E(a,\bar{a})}$ is a positive differential on R defining a Riemannian metric with Gauss curvature

$$\frac{4\theta^2(t)E^2(a,\bar{a})}{\theta^2(t-a+\bar{a})} \frac{\partial^2}{\partial a \partial \bar{a}} \ln \frac{\theta(t-a+\bar{a})}{E(a,\bar{a})} = - \frac{4\theta^3(t)\theta(t-2a+2\bar{a})}{\theta^4(t-a+\bar{a})} < 0$$

by (41), (125) and Cor. 6.13. This metric generalizes the Poincaré metric in the unit disc D since if $C \xrightarrow{p} \mathbb{P}_1(\mathbb{C})$ is a conformal homeomorphic with $p(R) = D$,

$$\frac{1}{iE(a,\bar{a})} = \frac{1}{i}\sqrt{dp(a)dp(\bar{a})} \Big/ \frac{1}{\overline{p(a)}} - p(a) = \frac{|dp(a)|}{1 - |p(a)|^2}$$

$\forall a \in R$. When $t \in \hat{T}_0$ is a half-period e, this metric comes from $2\pi\sigma_e(\bar{a},a)$ where $\sigma_e(\bar{x},y)$, the Szego reproducing kernel for sections of L_e, is given by $\sum_1^\infty \overline{\phi_j(x)}\ \phi_j(y)$ for a complete set of holomorphic sections of L_e on $R \cup \partial R$ orthonormalized by the conditions $\int_{\partial R} \phi_j \bar{\phi}_k = \delta_j^k$:

Proposition 6.14. For any (even) half-period $e \in T_0$, let $\sigma_e(\bar{x},y) = \frac{1}{2\pi i} \frac{\Theta[e](y-\bar{x})}{\Theta[e](0)E(y,\bar{x})}$. Then $\sigma_e(\bar{x},y)$ is holomorphic in $\bar{x}$ and y except for a pole along $y = \bar{x}$, and satisfies

$$\sigma_e(\bar{x},y) = -\sigma_e(y,\bar{x}) = -\overline{\sigma_e(x,\bar{y})} \qquad \forall\ x,y \in C.$$

For any section ϕ of L_e holomorphic on $R \cup \partial R$,

$$\phi(x) = \int_{\partial R} \sigma_e(\bar{y},x)\phi(y) = \int_{\partial R} \overline{\sigma_e(\bar{x},y)}\phi(y) \qquad \forall\ x \in R$$

so that $\sigma_e(\bar{x},y)$ is the Szego reproducing kernel for the space of holomorphic sections of L_e on $R \cup \partial R$ with the norm $\|\phi\| = (\int_{\partial R} |\phi|^2)^{\frac{1}{2}}$.

Proof. First observe that σ_e actually exists since $(\Theta)_{sing} \cap T_0 = \phi$ by Prop. 6.8; from the symmetry properties (127) and (131):

$$\overline{\sigma_e(\bar{x},y)} = -\frac{1}{2\pi i} \frac{\Theta[\phi(e)](\bar{y}-x)}{\Theta[\phi(e)](0)E(\bar{y},x)} = -\frac{1}{2\pi i} \frac{\Theta[e](x-\bar{y})}{\Theta[e](0)E(\bar{y},x)} = -\sigma_e(x,\bar{y}).$$

By Prop. 6.11, this means that in terms of the symmetric open covering $\{U_\alpha\}$, $\overline{\sigma_e(\bar{x},y)}_{\beta',\alpha} = -\sigma_e(x,\bar{y})_{\beta,\alpha'}$, where $\sigma_e(x,\bar{y})_{\beta,\alpha'}$ is the section σ on the open set $U_\beta \times U_{\alpha'}$; consequently, if $\phi_\alpha(y)$ is any section ϕ of L_e on U_α, $\overline{\sigma_e(\bar{x},y)}_{\beta',\alpha}\phi_\alpha(y)$ is a section of $|K_C|$ in y and of L_e in x with the property that

$$-\overline{\sigma_e(\bar{x},y)}_{\beta',\alpha}\phi_\alpha(y) = \sigma_e(x,\bar{y})_{\beta,\alpha'}\phi_\alpha(y) = \sigma_e(x,y)_{\beta,\alpha}\phi_\alpha(y)$$

if $y \in U_\alpha \cap \partial R$ for some boundary disc U_α. Therefore, if $x \in R$:

$$\int_{y\in\partial R} \overline{\sigma_e(\bar{x},y)}\phi(y) = \frac{1}{2\pi i}\int_{y\in\partial R} \frac{\Theta[e](y-x)}{\Theta[e](0)E(x,y)}\phi(y) = \operatorname*{Res}_{y=x} \frac{\Theta[e](y-x)\phi(y)}{\Theta[e](0)E(x,y)} = \phi(x).$$

In the case of a planar domain ($\rho = 0$), there is a global univalent function Z on R with dZ(x) a nowhere vanishing differential having a well-defined square root on R cut along segments joining $\Gamma_1,\ldots,\Gamma_{n-1}$ to Γ_0; then $\dfrac{\sqrt{dZ(x)}}{|dZ(x)|^{\frac{1}{2}}} = \exp \dfrac{i}{2} \operatorname{Arg} \dfrac{dZ(x)}{|dZ(x)|}$ is a multi-valued function on R which picks up a factor of (-1) as x traverses any loop Γ_k, $1 \le k \le n-1$. On the other hand, if $b \in \Gamma_0$, $\mu \in (\mathbb{Z}/2\mathbb{Z})^{n-1}$ and $e = \begin{Bmatrix} \frac{1}{2}\mu \\ 0 \end{Bmatrix}_\tau \in T_0 \cap S$, $\sigma_e(b,y)/|dZ(b)dZ(y)|^{\frac{1}{2}}$ is real with $1+\mu_k$ zeroes (mod 2) on Γ_k by Props. 6.4 and 6.8. Therefore by continuity, $\sigma_e(\bar{x},y)/\sqrt{dZ(\bar{x})dZ(y)}$ is a multiplicative function on $R \times R$ which picks up the factor $(-1)^{\mu_k}$ as x goes around the loop Γ_k. So when $e = \begin{Bmatrix} 0 \\ 0 \end{Bmatrix}_\tau$, the classical Szego kernel $\sigma_0(\bar{x},y)/\sqrt{dZ(\bar{x})dZ(y)}$ is well-defined on $R \times R$ and a reproducing kernel for a space of *functions* on R, while $\sigma_e(\bar{x},y)/\sqrt{dZ(\bar{x})dZ(y)}$ for $e \neq 0$ is a reproducing kernel for sections of $e \in J_0$ as given by (6) - that is, functions with multipliers $(-1)^{\mu_k}$ along Γ_k, $k = 1,\ldots,n-1$.

Now in the case when R is the unit disc D, the inner product on the holomorphic half-order differentials can also be obtained by integrating two analytic functions over ∂D with measure given by the inner normal derivative of the Green's function G(x,0) at $x \in \partial D$ for the basepoint $0 \in D$. To describe this situation in the general case, recall that on a finite surface R, if

$$(132) \quad \Omega_{b-a}(x) = \overline{\Omega_{\bar{b}-\bar{a}}(\bar{x})} = \omega_{b-a}(x) - \sum_{j,k=1}^{g} u_j(x)(\operatorname{Re}\tau)^{-1}_{jk} \operatorname{Re} \int_a^b u_k$$

is the unique differential of the third kind on C with simple poles of residue -1 and +1 at a and b respectively and with purely imaginary periods over all cycles on C, then the Green's function

$$G(x,y) = \tfrac{1}{2}\int_{\bar{x}}^{x} \Omega_{\bar{y}-y} = \tfrac{1}{2}\int_y^{\bar{y}} \int_{\bar{x}}^{x} \omega(p,q) + \tfrac{1}{2}\sum_{j,k=1}^{g} (\operatorname{Re}\tau)^{-1}_{jk} \int_x^{\bar{x}} \operatorname{Re} u_j \int_y^{\bar{y}} \operatorname{Re} u_k$$

is a harmonic function in x and y with the symmetries

$$G(x,y) = G(y,x) = -G(x,\bar{y}) = \overline{G(x,y)} \qquad \forall\, x,y \in C,$$

and with a local expansion at $y = x$:

$$(133) \quad G(x,y) = \ln \frac{1}{|x-y|} + \ln iE(x,\bar{x}) + \tfrac{1}{2}\sum_{1}^{g} (\operatorname{Re}\tau)_{jk} m_j(x) m_k(x) + O(|x-y|)$$

in terms of the harmonic measures of (125). The bilinear differential

$$B(x,y) = \frac{i}{\pi} \frac{\partial^2}{\partial x \partial \bar{y}} G(x,\bar{y})\, dx\, d\bar{y} = \frac{i}{2\pi}\left[\omega(x,\bar{y}) - \frac{1}{4}\sum_{1}^{g} (\operatorname{Re}\tau)^{-1}_{jk} \left(u_j(x) - \overline{u_j(\bar{x})}\right)\left(u_k(\bar{y}) - \overline{u_k(y)}\right)\right]$$

is the Bergman kernel of C with the reproducing property: $V(x) = \iint_R B(x,y) \wedge V(y)$ for any differential $V(x)$ holomorphic on $R \cup \partial R$.

<u>Proposition 6.15.</u> For any fixed $a \in R$, let $\Omega_{\bar{a}-a}(x) = dG(x,a) + i * dG(x,a)$ be the differential of the third kind on C with poles of residue $-1,+1$ at $a,\bar{a}$ and purely imaginary periods along all closed paths in C. If $\mathcal{A}$ is the divisor of zeroes of $\Omega_{\bar{a}-a}$ in R, then $i(\mathcal{A}) = 0$ and $e = \mathcal{A} - a - \Delta$ is a point of $T_{0,\ldots,0}$ satisfying

$$(134) \qquad \frac{\partial \ln \Theta}{\partial z_j}(e+a-\bar{a}) - \frac{\partial \ln \Theta}{\partial z_j}(e) = m_j(a) \qquad j = 1,\ldots,g$$

for m_j the harmonic measures of (125). For any $x,y \in C$, let $\Lambda_a(x,y)$ be the Cauchy kernel (37) formed from the divisor $\mathcal{A} \subset R$; then the meromorphic function of $x,\bar{y} \in C$:

$$K(x,\bar{y}) = \frac{\Lambda_a(x,\bar{y})}{\Omega_{\bar{a}-a}(x)} = \frac{\Theta(x-\bar{y}+e)}{\Theta(a-\bar{y}+e)}\,\frac{\Theta(a-\bar{a}+e)}{\Theta(x-\bar{a}+e)}\,\frac{E(\bar{y},a)}{E(x,\bar{y})}\,\frac{E(x,\bar{a})}{E(\bar{a},a)}$$

is, for any $y \in R \cup \partial R$, a holomorphic function of $x \in R$ such that

$$f(x) = \frac{i}{2\pi}\int_{y \in \partial R} f(y) K(x,\bar{y}) \Omega_{\bar{a}-a}(y) \qquad \forall\, x \in R$$

for all holomorphic functions f on $R \cup \partial R$. Thus $\overline{K(x,\bar{y})}$ is the reproducing kernel for the Hilbert space $H_2(R)$ of functions f analytic on R with finite norm $\|f\| = \lim_{\varepsilon\to 0^+} \left(-\frac{1}{2\pi}\int_{G(x,a)=\varepsilon} |f(x)|^2 * dG(x,a)\right)^{\frac{1}{2}}$.

Proof. The Green's function $G(x,a) > 0$ for all $x \in R$ so by the maximum principle, $0 < -*dG(x,a) = i\Omega_{\bar{a}-a}(x)$ for $x \in \partial R$; by Prop. 6.8, $e = \mathcal{A} - a - \Delta$ is therefore a point of $T_{0,\dots,0}$ with $i(\mathcal{A}) = 0$ and $e \notin (\Theta)$ since $(\Theta) \cap T_0 = \phi$. Since $\Omega_{\bar{a}-a}(x) = \dfrac{\Theta(x-a-e)\Theta(x-\bar{a}+e)E(a,\bar{a})}{\Theta(e)\Theta(a-\bar{a}+e)E(x,a)E(x,\bar{a})}$, Prop. 2.10 (38), (125) and (132) imply that $\forall\, x \in C$,

$$\sum_{i=1}^{g}\left[\frac{\partial \ln\theta}{\partial z_i}(e) - \frac{\partial \ln\theta}{\partial z_i}(e+a-\bar{a})\right]u_i(x) = -\sum_{i,k=1}^{g}(\mathrm{Re}\,T)^{-1}_{ik}\,\mathrm{Re}\left(\int_a^{\bar{a}} u_k\right)u_i(x) = -\sum_{i=1}^{g} m_i(a)u_i(x)$$

which gives (134). Now $\Theta(p-\bar{a}+e) = \overline{\Theta(\bar{p}-a-e)}$ never vanishes for $p \in R$, and hence $K(x,\bar{y})$ is holomorphic $\forall\, x,y \in R$ and $K(x,y)\Omega_{\bar{a}-a}(y)$ is holomorphic for $x,y \in R$ except for a simple pole at $y = x$. So if f is holomorphic on $R \cup \partial R$, the residue theorem gives

$$\frac{i}{2\pi}\int_{y\in\partial R} f(y)K(x,\bar{y})\Omega_{\bar{a}-a}(y) = -\frac{1}{2\pi i}\int_{y\in\partial R} f(y)K(x,y)\Omega_{\bar{a}-a}(y)$$

$$= -\operatorname*{Res}_{y=x} f(y)\,\Lambda_a(x,y)\,\frac{\Omega_{\bar{a}-a}(y)}{\Omega_{\bar{a}-a}(x)} = f(x) \qquad \forall\, x \in R.$$

The reproducing property of $\overline{K(x,\bar{y})}$ for the Hilbert space $H_2(R)$ is then a consequence of the general Poisson representation formula for functions in the Hardy class $H_1(R)$.

Planar Domains. For the remainder of this chapter, we assume that C of genus $g = n-1$ is the double of a planar domain bordered by n analytic curves $\Gamma_0,\dots,\Gamma_{n-1}$.

Proposition 6.16. For all $s \in \hat{S}_0$, $\Theta(s) > 0$ and $\Theta(x-a-s)$ has g zeroes in $\bar{R}$ for any $a \in R$. Every unitary function on C, holomorphic on R with the minimal (g+1) number of zeroes has the form $\varepsilon \dfrac{\Theta(x-\bar{a}-s)\ E(x,a)}{\Theta(x-a-s)\ E(x,\bar{a})}$ for $s \in \hat{S}_0$, $a \in R$ and $|\varepsilon| = 1$.

Proof. If $s \in \hat{S}_0 \cap (\Theta)$, then by Cor. 6.5, $s = \zeta-\Delta$ where ζ is positive of degree $g-1$ with an odd number of points on $\Gamma_1,\ldots,\Gamma_{n-1}$ - an impossibility since $g-1 < n-1$; thus $(\Theta) \cap \hat{S}_0 = \phi$ and $\Theta(s) > 0$ for all $s \in \hat{S}_0$ since $\Theta(s)$ is real on $\hat{S}_0$ by Prop. 6.1 and $\Theta(0) > 0$ by Cor. 6.13. Now if $\Theta(b-a-s) = 0$ for some $b \in \Gamma_k$ and $a \in R$, $-s = D+\bar{a}+a-b-\Delta$ where, by Prop. 6.4, D is positive with an odd number of points on each contour except Γ_k (and possibly Γ_0); again this is impossible since $\deg D = g-2 < n-2$. Thus as a varies over the interior of R, $\operatorname{div}_C\Theta(x-a-s)$ has a fixed number of, say, d points in $\bar{R}$ and $g-d$ points in R. To compute d, let $a \in R$ approach a point $b \in \Gamma_0$ and set $\operatorname{div}_C\Theta(x-a-s) = \sum_1^g a_k$ and $\operatorname{div}_C\Theta(x-b-s) = \sum_1^g b_k$ where $b_k \in \Gamma_k$ by Prop. 6.4. If δ is a local coordinate of a near b, and ε_k a local coordinate for a_k near b_k, then the condition $\int_b^a u = \sum_1^g \int_{b_k}^{a_k} u$ implies that $\delta u(b) = u(\mathfrak{B})\varepsilon$ where $u(\mathfrak{B})$ is the non-singular $g\times g$ matrix $(u_i(b_j))$. From (35) of Lemma 2.7,

$$\varepsilon_k = \delta \sum_1^g u(\mathfrak{B})^{-1}_{kj}\ u(b)_j = \delta \lim_{x\to b_k} E(x,b_k)^2\ \frac{\partial^2 \ln \Theta(x-b-s)}{\partial x\partial b} = \delta\ \frac{H_f(b)}{H_f(b_k)}$$

where $f = b_k-b-s \in (\Theta) \cap S_{0,\ldots,\overset{(k)}{1},\ldots,0}$ and $H_f(x) = \sum_1^g \frac{\partial\Theta}{\partial z_i}(f)u_i(x)$ is a differential with a symmetric divisor of zeroes on C. However,

$$H_f(b)\overline{H_f(b_k)} = H_f(b)H_{\phi(f)}(b_k) = H_f(b)H_f(b_k)\exp\{-\tfrac{1}{2}\tau_{kk} - \int_b^{b_k} u_k + s_k\}$$

$$= -E^{-1}(b,b_k)\overline{E^{-1}(b,b_k)}\Theta(s)\Theta\left(s + 2b - 2b_k - \begin{Bmatrix}0 & \ldots & \overset{(k)}{1} & \ldots & 0\\ 0 & \ldots & 0 & \ldots & 0\end{Bmatrix}_\tau\right)e^{-\frac{1}{2}\tau_{kk}}$$

by (20), (25), (127) and (131). Since $s+2b-2b_k - \begin{Bmatrix} 0 & \dots & 1 & \dots & 0 \\ 0 & \dots & 0 & \dots & 0 \end{Bmatrix}_\tau \in \hat{S}_0$ (with the 1 in the (k) position)

if $s \in \hat{S}_0$, we conclude that $\varepsilon_k/\delta < 0$ for $k = 1,\dots,n-1$; and so all zeroes of $\Theta(x-a-s)$ lie in $\bar{R}$ for $a \in R$ near b, and by continuity $d = \deg \operatorname{div}_{\bar{R}}\Theta(x-a-s) = g$ for all $a \in R$. Finally suppose that f is a unitary function on C such that $\operatorname{div}_C f = \bar{\mathcal{A}} + a - \mathcal{A} - \bar{a}$ with $a \in R$ and $\mathcal{A} = \sum_1^N a_j$ contained in $\bar{R}$; then the harmonic measures $m_i(a) > 0$ and $m_i(\bar{a}_j) > 0$ of (125) satisfy

$$\sum_{k=1}^{g} \tau_{ik}^{-1} \int_{\mathcal{A}+\bar{a}}^{\bar{\mathcal{A}}+a} u_k = m_i(a) + \sum_{j=1}^{N} m_i(\bar{a}_j) = M_i \geq 1 \qquad \text{for } M_i \in \mathbb{Z}.$$

But $\sum_1^{n-1} m_i(x) + m_0(x) \equiv 1 \quad \forall\, x \in R$, where m_0, the harmonic measure of R with respect to Γ_0, satisfies a similar condition $m_0(a) + \sum_{j=1}^{N} m_0(\bar{a}_j) = M_0 \geq 1$, $M_0 \in \mathbb{Z}$. Therefore

$$n \leq \sum_{i=0}^{n-1} M_i = \sum_{i=0}^{n-1} \{m_i(a) + \sum_{j=1}^{N} m_i(\bar{a}_j)\} = N + 1 \quad *$$

and any unitary function holomorphic on R must have at least $n = g+1$ zeroes. Furthermore, if the function has exactly n zeroes, equality must hold in the above inequalities, and this means that $s = \int_{ga}^{\mathcal{A}} u - k^a \in \mathbb{C}^g$ must be in $\hat{S}_0$ since, by (129):

$$\phi(s) - s = \int_{\mathcal{A}+\bar{a}}^{\bar{\mathcal{A}}+a} u + k^b - \phi(k^b) = \begin{Bmatrix} M_1 & \dots & M_g \\ 0 & \dots & 0 \end{Bmatrix}_\tau - \begin{Bmatrix} 1 & \dots & 1 \\ 0 & \dots & 0 \end{Bmatrix}_\tau = 0 \quad \underline{\text{in } \mathbb{C}^g}$$

for any $b \in \Gamma_0$.

Using this result, a solution can be given to an extremal problem for bounded analytic functions as formulated in [3, p. 123]:

---

* See [2, p. 7]; the inequality $n \leq N+1$ of course holds for arbitrary bordered surfaces by the argument principle.

Proposition 6.17. For two distinct fixed points a and b in R, let $\mathcal{W}$ be the family of all differentials ω analytic on $R \sqcup \partial R$ except for simple poles at a and b with $\operatorname{Res}_{x=b} \omega(x) = 1$, and denote by $\mathfrak{F}$ the family of functions F vanishing at a and analytic on $R \sqcup \partial R$ where $|F| \leq 1$. Then

$$|F(b)| \leq \frac{1}{2\pi} \int_{\partial R} |\omega| \qquad \forall\, \omega \in \mathcal{W}, \quad F \in \mathfrak{F}$$

with equality attained if and only if, for $s = \frac{1}{2}\int_{a+\bar{a}}^{b+\bar{b}} u \in \hat{S}_0$,

$$\omega(x) = \frac{\theta^2(x-a-s)}{\theta^2(b-a-s)} \frac{E(x,\bar{a})E(b,\bar{b})E(a,b)}{E(x,b)E(x,\bar{b})E(x,a)E(b,\bar{a})},$$

$$F(x) = \varepsilon\, \frac{\theta(x-\bar{a}-s)}{\theta(x-a-s)} \frac{E(x,a)}{E(x,\bar{a})}, \quad |\varepsilon| = 1 \quad \text{and} \quad |F(b)| = \frac{\theta\left(\frac{1}{2}\int_{\bar{a}+\bar{b}}^{a+b} u\right)}{\theta\left(\frac{1}{2}\int_{b+\bar{a}}^{a+\bar{b}} u\right)} \left|\frac{E(b,a)}{E(b,\bar{a})}\right|.$$

Proof. We will find the extremal F and ω, assuming their existence; the explicit construction will show they satisfy the required properties. Now by Cauchy's Theorem, $|F(b)| = \frac{1}{2\pi}\left|\int_{\partial R} F\omega\right| \leq \frac{1}{2\pi}\int_{\partial R} |\omega|$ with equality if and only if $F\omega = \varepsilon_1|\omega|$ on ∂R for some constant ε_1 of absolute value 1; thus $|F| = 1$ on ∂R and F extends to a unitary function on C with $\operatorname{div}_C F = a + \bar{\mathcal{A}} - \bar{a} - \mathcal{A}$ where $\bar{\mathcal{A}} \subset R$ and $\deg \mathcal{A} \geq g$ by Prop. 6.16. On the other hand, $\frac{1}{\varepsilon_1}F(x)\omega(x) = \psi(x)$ is a positive differential on ∂R and so extends to a symmetric differential on C with $\operatorname{div}_{R \sqcup \partial R} \psi = \bar{\mathcal{A}} + a + \operatorname{div}_{R \sqcup \partial R} \omega$; since ω on R has poles only at a and b, and since $\deg \operatorname{div}_C \psi = 2g-2$, we conclude that ω actually has no zeroes on $R \sqcup \partial R$ and that $\deg \mathcal{A} = g$. By Prop. 6.16 then,

$$F(x) = \varepsilon\, \frac{\theta(x-\bar{a}-s)E(x,a)}{\theta(x-a-s)E(x,\bar{a})} \quad \text{with} \quad |\varepsilon| = 1, \quad s = \mathcal{A} - a - \Delta \in \hat{S}_0$$

and by Cor. 6.9,

$$\psi(x) = r\,\frac{\Theta(x-b-t)\Theta(x-\bar{b}+t)}{E(x,b)E(x,\bar{b})} \quad \text{with} \quad r \in \mathbb{R}^+, \quad t = \mathcal{A}-b-\Delta \in \hat{T}_0.$$

Since $S_0 \oplus T_0 = J(C)$ and

$$s + t = s - \phi(t) = \bar{b} - a = \tfrac{1}{2}\int_{a+\bar{a}}^{b+\bar{b}} u + \tfrac{1}{2}\int_{a+b}^{\bar{a}+\bar{b}} u \in J(C),$$

we find $s = \tfrac{1}{2}\int_{a+\bar{a}}^{b+\bar{b}} u$ and $\omega(x) = \varepsilon_1 \frac{\psi(x)}{F(x)} = \frac{\varepsilon_1 r}{\varepsilon}\,\frac{\Theta(x-a-s)^2 E(x,\bar{a})}{E(x,a)E(x,b)E(x,\bar{b})}$

where $\frac{\varepsilon}{\varepsilon_1 r} = \operatorname*{Res}_{x=b} \frac{\Theta(x-a-s)^2 E(x,\bar{a})}{E(x,a)E(x,b)E(x,\bar{b})} = \frac{\Theta\left(\tfrac{1}{2}\int_{\bar{a}+b}^{a+\bar{b}} u\right)^2 E(\bar{a},b)}{E(b,a)E(b,\bar{b})}$.

Letting $b \to a$, the above proof gives:

Corollary 6.18 (Schwarz' Lemma). For $a \in R$ fixed, let ω be any differential analytic on $R \cup \partial R$ except at a where $\omega(x) - \frac{dz(x)}{(z(x)-z(a))^2}$ is holomorphic for some local coordinate z in a neighborhood of $x = a$; and let F be a function analytic on $R \cup \partial R$ where $|F| \leq 1$, and vanishing at a with Taylor development $F(x) = F'(a)(z(x)-z(a)) + \ldots$ in the same local coordinate z. Then $|F'(a)| \leq \frac{1}{2\pi}\int_{\partial R} |\omega|$, with equality if and only if

$$F(x) = \varepsilon\,\frac{\Theta(x-\bar{a})}{\Theta(x-a)}\,\frac{E(x,a)}{E(x,\bar{a})}, \qquad |\varepsilon| = 1$$

$$\omega(x) = \frac{\Theta^2(x-a)}{\Theta^2(0)E^2(x,a)}, \quad \text{and} \quad |F'(a)| = \frac{\Theta(a-\bar{a})}{i\Theta(0)E(a,\bar{a})}.$$

Observe that the extremal derivative $\frac{i}{\varepsilon}F'(a)$ is the positive differential of Cor. 6.13; also, the extremal function F(x) is $\varepsilon\,\frac{\sigma_0(x,\bar{a})}{\sigma_0(x,a)}$ for σ_0 the Szego-kernel of Prop. 6.14, a fact observed by Garabedian in [11, p. 22]. For a relation with the span of R, see [36, pp. 97-107].

As an example, let R be the annulus $\frac{1}{r} < |p| < 1$ and $\bar{R}$ the annulus $1 < |p| < r$ for $p \in \mathbb{C}$. Then identifying $|p| = \frac{1}{r}$ and $|p| = r$ under the anti-conformal involution $\phi(p) = \frac{1}{\bar{p}}$, $C = R \cup \partial R \cup \phi(R)$ is a compact surface of genus 1 with normalized differentials $u(p) = \frac{dp}{p} = \overline{-u(\phi(p))}$ and period matrix $\tau = \int_r^{1/r} u = -2\ln r < 0$ with respect to the canonical homology basis $A = \{re^{i\Theta},\ 0 \leq \Theta \leq 2\pi\}$ and $B = \{r-p,\ 0 \leq p \leq r - \frac{1}{r}\}$. The Riemann divisor class $\Delta = \pi i - \ln r$, and for $a \in C$, $\theta(\int_a^x u - z) = 0$ if and only if $\ln \frac{-rx}{a} = z$ in J_0.

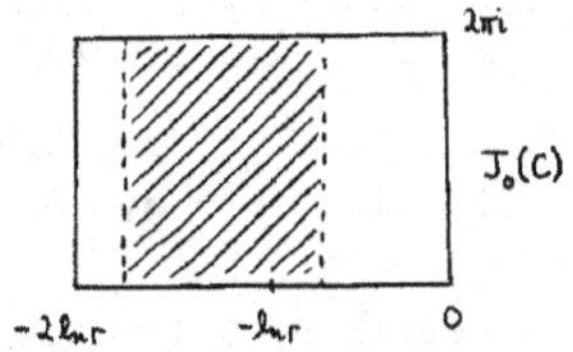

Thus for fixed $a \in R$, $\operatorname{div}_C \theta(\int_a^x u - z)$ is contained in R iff z is a point in a half of the fundamental parallelogram defined by $-\ln r^2|a| < \operatorname{Re} z < -\ln r|a|$. The variety S is the union of two circles S_0 and S_1 defined, respectively by $\operatorname{Re} z = 0$ and $\operatorname{Re} z = -\ln r$ in J_0; for $a \in R$ fixed, $\operatorname{div}_C \theta(\int_a^x u - s) = -are^{2\pi i\nu} \in \phi(R)$ for $s = \begin{Bmatrix} 0 \\ \nu \end{Bmatrix}_\tau \in S_0$ $(\nu \in \mathbb{R})$, while $\operatorname{div}_C \theta(\int_a^x u - s) = -ae^{2\pi i\nu} \in R$ for $s = \begin{Bmatrix} \frac{1}{2} \\ \nu \end{Bmatrix}_\tau \in S_1$. The variety T is the union of two circles T_0 and T_1 defined by $\operatorname{Im} z = 0$ and $\operatorname{Im} z = \pi$ in J_0; for any $a \in C$, $\arg \operatorname{div}_C \theta(\int_a^x u - t) = \arg a$ (resp. $\pi + \arg a$) for $t \in T_1$ (resp. T_0). The Weierstrass functions for the lattice generated by $2\pi i$ and $-2\ln r$ have the symmetries $\wp(z) = \overline{\wp(\bar{z})} = \wp(-z)$ and $\zeta(z) = \overline{\zeta(\bar{z})} = -\zeta(-z)$ $\forall z \in \mathbb{C}$; the period η given in (46) is real and so, for any $a \in R$, the elliptic function

$$h(z) = \zeta(\ln|a|^2 + \pi i + \ln r + z) - \zeta(\pi i + \ln r + z) + \ln|a|^2(\eta - \frac{1}{2\ln r})$$

is real and continuous for z real. Since $\int_0^{-2\ln r} h(z)dz = 0$ by definition of η, and since $h'(z)$ has zeroes in T_0 at $z = -\ln|a|$ and $-\ln r|a|$, we conclude that there is exactly one zero of $h(z)$ in the interval $(-\ln r|a|, -\ln|a|) \subset \mathbb{R}$, and this is the point in T_0 giving the reproducing kernel for $H_2(R)$ in Prop. 6.15. The Bergman kernel function, on the other hand, is given by

$$\frac{i}{2\pi}[\omega(x,\phi(y)) - (\operatorname{Re}\tau)^{-1}u(x)u(\phi(y))] = \frac{1}{2\pi i}\left(\wp(\ln x\bar{y}) - \eta + \frac{1}{2\ln r}\right)\frac{dx\,d\bar{y}}{x\bar{y}}$$

$\forall\ x,y \in C$, and the Szego kernel function for the half-period $\begin{Bmatrix}0\\0\end{Bmatrix}_\tau$ is

$$\sigma_0(\phi(y),x) = \frac{1}{2\pi i}\,\frac{\Theta\left(\int_x^{\phi(y)} u\right)}{\Theta(0)E(x,\phi(y))} = -\frac{1}{2\pi}\,\frac{\Theta(\ln x\bar{y})\Theta\begin{bmatrix}1\\1\end{bmatrix}'(0)}{\Theta(0)\Theta\begin{bmatrix}1\\1\end{bmatrix}(\ln x\bar{y})}\,\frac{\sqrt{dx\,d\bar{y}}}{(x\bar{y})^{\frac{1}{2}}}$$

$$= -\frac{1}{2\pi}\,\frac{\sum_{-\infty}^{+\infty} r^{-n^2}(x\bar{y})^n \sum_{-\infty}^{+\infty}(n+\frac{1}{2})r^{-(n+\frac{1}{2})^2}(-1)^{n+\frac{1}{2}}}{\sum_{-\infty}^{+\infty} r^{-n^2}\sum_{-\infty}^{+\infty} r^{-(n+\frac{1}{2})^2}(-x\bar{y})^{n+\frac{1}{2}}}\,\frac{\sqrt{dx\,d\bar{y}}}{(x\bar{y})^{\frac{1}{2}}}$$

$\forall\ x,y \in C$; $\sigma_0(\phi(y),x)$ is zero when $x\bar{y} = -r$ and is infinite when $x\bar{y} = 1$. Observe that as $r \to +\infty$, $\sigma_0(\phi(y),x)$ becomes the Szego-kernel $\frac{1}{2\pi}\frac{\sqrt{dx\,d\bar{y}}}{1 - x\bar{y}}$ for the unit disc D, $|x| < 1$, $x \in ₵$; and Cor. 6.18 implies that if $|F| \le 1$ in D and $F(a) = 0$ for some $a \in D$, then $|F'(a)| \le 2\pi\,\frac{\sigma_0(\phi(a),a)}{|da|} = \frac{1}{1-|a|^2}$, the classical Schwarz lemma.

Notation

τ	period matrix, 1 and 3
$\mathfrak{H}_g$	Siegel half-plane, 1
$\left\{\begin{matrix}\delta\\ \varepsilon\end{matrix}\right\}_\tau$	point with characteristics $\begin{bmatrix}\delta\\ \varepsilon\end{bmatrix}$, 1
$\theta\begin{bmatrix}\delta\\ \varepsilon\end{bmatrix}$	first order theta-function with characteristics $\begin{bmatrix}\delta\\ \varepsilon\end{bmatrix}$, 1
v_i	normalized holomorphic differentials, 3
$J(C) = J_0(C)$	Jacobi-Picard variety, 3
$\omega_{\mathfrak{B}-\mathfrak{A}}$, ω_{b-a}	normalized differentials of the third kind, 4
$J_n(C)$	components of the divisor class group, 5
$H^0(D)$	holomorphic sections of the line bundle with divisor class D, 5
K_C	canonical point in J_{2g-2}, 6
Δ	Riemann's divisor class, 7
k^b	Riemann constants with basepoint b, 8
$H_f(x)$	differential formed from gradient of (θ) at f, 10
L_α	bundles of half-order differentials, 11
$E(x,y)$	prime form , 16
$S(x)$	projective connection, 19
$\omega(x,y)$	bilinear differential of the second kind, 20
$\mathcal{O}_C$	sheaf of germs of holomorphic functions
Ω^1_C	sheaf of holomorphic differentials on C
$i(D)$	index of speciality of a divisor (class) D
$\mathrm{div}_C f(x)$	divisor of zeroes - poles of f on C

See also the notational convention on p. 5.

REFERENCES

[1] Accola, R., Vanishing properties of theta functions for Abelian covers of Riemann surfaces I and II. Brown University Notes, 1966.

[2] Ahlfors, L., Bounded analytic functions. Duke Math. Jour., Vol. 14, 1947.

[3] Ahlfors, L., Open Riemann surfaces and extremal problems on compact subregions. Comm. Math. Helv., Vol. 24, 1950.

[4] Andreotti, A., On a theorem of Torelli. Amer. Jour. of Math., Vol. 80, 1958.

[5] Andreotti, A. and Mayer, A., On period relations for Abelian integrals on algebraic curves. Scuola Normale Superiore - Pisa, 1967.

[6] Atiyah, M., Riemann surfaces and spin structures. Annales Scientifiques de L'École Normale Supérieure 1971, Vol. 4.

[7] Baker, H., Abel's theorem and the theory of theta functions. Cambridge 1897.

[8] Coble, A., Algebraic geometry and theta functions. A.M.S. Colloq. Pub., Vol. 10.

[9] Conforto, F., Abelsche funktionen und algebraische geometrie. Springer 1956.

[10] Farkas, H. and Rauch, H., Period relations of Schottky type on Riemann surfaces. Annals of Math., Vol. 92, 1970.

[11] Garabedian, P., Schwarz's lemma and the Szegö kernel function. Trans. A.M.S., Vol. 67, 1949.

[12] Grauert, H., Ein theorem der analytischen Garbentheorie. I.H.E.S. Pub. No. 5, 1960.

[13] Gunning, R., Lectures on Riemann surfaces. Princeton Math. Notes, 1966.

[14] Gunning, R., Lectures on vector bundles on Riemann surfaces. Princeton Math. Notes, 1967.

[15] Hawley, N. and Schiffer, M., Half-order differentials on Riemann surfaces. Acta Math., Vol. 115, 1966.

[16] Hurwitz, A. and Courant, R., Funktionentheorie. Springer, 1964 edition.

[17] Igusa, J., Theta functions. Springer, 1972.

[18] Klein, F., Zur theorie der Abel'schen functionen. Math. Ann., Vol. 36, 1890.

[19] Krazer, A., Lehrbuch der thetafunktionen. Chelsea, 1970.

[20] Krazer, A. and Wirtinger, W., Abelsche funktionen und allgemeine theta-funktionen. Enzykl. der Math. Wissen., II B 7, pp. 604-873.

[21] Lewittes, J., Riemann surfaces and the theta functions. Acta Math., Vol. 111, 1964.

[22] Mayer, A., On the Jacobi inversion theorem. Princeton Thesis, 1961.

[23] Mumford, D., On the equations defining Abelian varieties I. Inven. Math., Vol. 1, 1966.

Mumford, D., Notes on the Prym variety and Schottky relations. (Unpublished).

[24] Poincaré, H., Remarques diverses sur les fonctions abéliennes. Jour. de Math. (5th series), Vol. 1, 1895.

[25] Rauch, H., Weierstrass points, branch points and moduli of Riemann surfaces. Comm. Pure and Applied Math., Vol. 12, 1959.

[26] Recillas, S., A relation between curves of genus 3 and curves of genus 4. Brandeis Univ. thesis, 1970.

[27] Riemann, B., Collected works, Nachträge IV. Dover, 1953.

[28] Schottky, F., Uber die moduln der thetafunctionen. Acta Math., Vol. 27, 1903.

[29] Schottky, F. and Jung, H., Neue Sätze über Symmetralfunctionen und die Abel'schen Functionen der Riemann'schen Theorie: Erste und Zweite Mittheilung. Berlin Akad. der Wissen., Sitzungsberichte 1909, Vol. 1.

[30] Stahl, H., Abel'sche functionen. Teubner, 1896.

[31] Thomae, J., Beitrag zur Bestimmung von $\theta(0)$ durch die Klassenmoduln algebraischer Functionen. Crelle's Jour., Vol. 71, 1870.

[32] Weil, A., Variétés Abéliennes et Courbes Algebriques. Hermann, 1948.

[33] Weil, A., Variétés Kahlériennes. Hermann, 1958.

[34] Wirtinger, W., Untersuchungen über Thetafunctionen. Teubner, 1895.

[35] Seminar on degeneration of algebraic varieties. (I. A. S.) Fall 1970.

[36] Hejhal, D., Theta functions, kernel functions and Abelian integrals. A.M.S. Memoir 129, 1972.

[37] Gunning, R., Lectures on Jacobi varieties. Princeton Math. Notes, 1972.